Engineering Secure Two-Party Computation Protocols

Thomas Schneider

Engineering Secure Two-Party Computation Protocols

Design, Optimization, and Applications
of Efficient Secure Function Evaluation

Springer

Thomas Schneider
European Center for Security
  and Privacy by Design (EC-SPRIDE)
Technische Universität Darmstadt
Darmstadt
Germany

ISBN 978-3-642-30041-7   ISBN 978-3-642-30042-4   (eBook)
DOI 10.1007/978-3-642-30042-4
Springer Heidelberg New York Dordrecht London

Library of Congress Control Number: 2012940722

Printed on acid-free paper

Springer is part of Springer Science+Business Media (www.springer.com)

To my family

Preface

Secure two-party computation, called Secure Function Evaluation (SFE), enables two mutually mistrusting parties (client & server) to evaluate an arbitrary function f on their respective private inputs x, y while revealing nothing but the result $z = f(x; y)$. Since its invention by Andrew Chi-Chih Yao in the 1980s, SFE has gained the attention of many researchers in cryptography, but was widely believed to be too inefficient for practical privacy-preserving applications. In recent years, the rapidly growing speed of computers and communication networks, algorithmic improvements, automatic generation, and optimizations of SFE protocols have made them usable in many practical application scenarios.

Engineering such efficient SFE protocols for practical privacy-preserving applications is a rapidly emerging topic in many top conferences on information security and applied cryptography and several research projects are working on this subject, e.g., the EU funded project "Computer Aided Cryptography Engineering" (CACE) or the U.S. funded project "Programming on Encrypted Data" (PROCEED). Also, for the first time companies are adapting SFE technology for their products.

In contrast to previous books, we cover the practical aspects of SFE protocols, in particular their systems and implementation aspects. Hence, this book is an ideal counterpart to Hazay and Lindell's book "Efficient Secure Two-Party Protocols: Techniques and Constructions" (Springer 2010) which focuses on the underlying security definitions, general constructions for different adversary models, and constructions for specific two-party functionalities. While Hazay and Lindell consider the (theoretical) efficiency of protocols and stronger covert and malicious adversaries, we concentrate on today's best performing constructions for the semi-honest adversaries setting.

We present advanced state-of-the-art techniques in the design, optimization, and applications of (practically) efficient SFE protocols. This makes the book essential for researchers, students, and practitioners in the area of applied cryptography and information security who aim to construct practical cryptographic protocols for privacy-preserving real-world applications.

Outline. This book contains the following main parts:

- *Basics of Practically Efficient Secure Function Evaluation*. We give a detailed overview on state-of-the-art techniques and optimizations for efficient Secure Function Evaluation (SFE) with special focus on their practicability. In particular, we are concerned with their performance and possibilities for pre-computations.
- *Circuit Optimizations and Constructions*. The complexity of today's most efficient SFE protocols depends linearly on the size of the Boolean circuit representation of the evaluated function. Further, recent techniques for SFE based on so-called improved Garbled Circuits (GCs) allow for very efficient secure evaluation of XOR gates. We give transformations that substantially reduce the size of Boolean circuits if the costs for evaluating XOR gates are much lower than for other types of gates. Our optimizations provide more efficient circuits for standard functionalities such as integer comparison, finding minima/maxima, and fast multiplication. Example applications that benefit from our improvements include secure first-price auctions.
- *Hardware-Assisted GC Protocols*. We improve the deployability of SFE protocols by using tamper-proof Hardware (HW) tokens. In particular, GCs can be generated by a tamper-proof HW token which is provided by the server to a client but not trusted by the client. The presented HW-assisted SFE protocol makes the communication between client and server independent of the size of the evaluated function. Further, we show how GCs can be evaluated in HW in a leakage resilient way, so-called One-Time Programs. As an application we show how the combination of GCs and tamper-proof HW allows one to securely outsource data to an untrusted cloud service provider such that arbitrary functions can be computed securely on the data with low latency.
- *Modular Design of Efficient SFE Protocols*. Automatic generation of SFE protocols from high-level specifications makes SFE usable for application programmers and yields less-error-prone implementations. We present a framework which enables one to modularly design efficient SFE protocolsas a sequence of operations on encrypted data. In our framework, efficient SFE protocols based on Homomorphic Encryption and GCs can be combined while abstracting from the underlying cryptographic details. Our corresponding language and tool, called Tool for Automating Secure Two-party Computations (TASTY), allow one to describe, automatically generate, execute, and benchmark such modular and efficient SFE protocols. As an application example we consider privacy-preserving face recognition.

Comments and Errata. Your feedback on the book or any errors you may find are highly appreciated. Please e-mail your comments and errata to thomaschneider@gmail.com. A list of known errata will be maintained at http://thomaschneider.de/engineeringSFEbook.

Acknowledgments

This book is a revised and extended version of my Ph.D. thesis, written at the Ruhr-University Bochum. Hence, first of all I would like to thank my Ph.D. advisors Ahmad-Reza Sadeghi and Benny Pinkas. I am thankful to Ahmad for his outstanding supervision and the opportunities he opened up for me. His mentorship in the form of a mixture of giving initial directions and topics to look into, the freedom to develop my own research ideas, and a lot of valuable feedback was just perfect. He showed me that our research community is like a big family and gave me lots of chances to talk to, learn from, work with, and spend a great time together with many outstanding fellow researchers. I am very honored to have Benny as external advisor and thank him a lot for our productive discussions on deep technical details and his hospitality.

Thanks also to all my co-authors beyond Ahmad and Benny with whom I had the pleasure to collaborate in different areas of cryptography and security: I learned a lot from experienced senior researchers in secure computation, efficient implementations, zero-knowledge, signal processing, and formal verification (Endre Bangerter, Manuel Barbosa, Mauro Barni, Jan Camenisch, Marc Fischlin, Kimmo Järvinen, Vladimir Kolesnikov, Nigel P. Smart, Joe-Kai Tsay, Ivan Visconti). Thanks to my fellow Ph.D. students (José Bacelar Almeida, Stefania Barzan, Thomas Briner, Pierluigi Failla, Stephan Krenn, Riccardo Lazzeretti, Stephen C. Williams, Marcel Winandy) for the splendid exchange of mutually stimulating ideas. Special thanks also to the undergraduate students whom I was pleased to supervise and work with (Wilko Henecka, Stefan Kögl, Annika Paus, Immo Wehrenberg).

Many thanks also to all my colleagues at the System Security Lab at the Horst Görtz Institute for IT-Security in Bochum and the Center for Advanced Security Research Darmstadt for their support and many interesting discussions during lunches and coffee breaks.

Thanks also to Wilko Henecka, Kimmo Järvinen, Berry Schoenmakers, and my family for proofreading drafts of this book, and to my editor Ronan Nugent and the copyeditor for their support during the publishing process.

I am grateful for the inspiring lectures, exercises, summer schools, and mentorship of various people who awoke my interest in applied crypto during my studies: Falko Dressler, Vladimir Kolesnikov, Bernd Meyer, Helmut Meyn, Wolfgang M. Ruppert, Thorsten Schütze, Volker Strehl, and Susanne Wetzel.

Last but not least, I'd like to thank my wonderful wife Karolin for all her support and understanding while writing this book during the years we spent in Bochum and Darmstadt, the friends we made there, and our families and old friends for staying in touch. Even though lots of counting is involved while engineering cryptographic protocols, Albert Einstein brought it to the point:

Everything that can be counted does not necessarily count.
Everything that counts cannot necessarily be counted.

April 2011   Thomas Schneider

Contents

Abbreviations

AE	Algorithm Engineering
AES	Advanced Encryption Standard
CACE	Computer-Aided Cryptography Engineering, http://cace-project.eu.
DAG	Directed Acyclic Graph
EC	Elliptic Curve
ECG	Electro Cardiogram
ElGamal	ElGamal Cryptosystem [75]
FPGA	Field-Programmable Gate Array
GC	Garbled Circuit
HE	Homomorphic Encryption
HW	Hardware
IPSec	Internet Protocol Security
MAC	Message Authentication Code
OBDD	Ordered Binary Decision Diagram
ODBS	Oblivious Database Search
OPRF	Oblivious Pseudo-Random Function
OT	Oblivious Transfer
OTM	One-Time Memory
OTP	One-Time Program
PET	Privacy-Enhancing Technology
PKI	Public-Key Infrastructure
PRF	Pseudo-Random Function
RAM	Random Access Memory
RO	Random Oracle
RSA	RSA Cryptosystem [183]
SFE	Secure Function Evaluation
SHA	Secure Hash Algorithm
SMPC	Secure Multi-party Computation
SOPC	System on a Programmable Chip
SW	Software
TASTY	Tool for Automating Secure Two-party Computations

TASTYL	TASTY Input Language
TCG	Trusted Computing Group
TLS	Transport Layer Security
TPM	Trusted Platform Module
TTP	Trusted Third Party
UC	Universal Composability
UCi	Universal Circuit
VHDL	Very High Speed Integrated Circuit Hardware Description Language

Chapter 1
Introduction

1.1 Privacy-Enhancing Technologies

As today's world gets more and more connected, in many application scenarios
actors with different and potentially conflicting interests want to interact. Examples
are citizens and governments (electronic passport and electronic id), patients and
health insurers (electronic health card, e-health services), or companies and service
providers (cloud computing). In this context, it is of foremost importance that the
underlying IT systems and algorithms can fulfill the diverse security and privacy
requirements of the involved parties. In particular, if sensitive (e.g., medical) data
is processed by not fully trusted service providers (e.g., "in the cloud"), conformity
with data privacy protection laws must be guaranteed.

To protect privacy and avoid data leakage and misuse by insiders, several security mechanisms, often called Privacy-Enhancing Technologies (PETs), exist. We
summarize some of them in Table 1.1 and describe them in the following. When constructing practical privacy-preserving systems, a combination of these technologies
can be used to provide the most appropriate level of security, trust, and performance
for each part of the system.

Secure Channels. It is well-known that cryptography provides efficient solutions
to transfer sensitive data over untrusted communication channels such as the Internet.
Such secure channels can be established based on a combination of (1) authentication
to ensure that the intended parties are communicating with each other and (2) key
exchange to establish keys to (3) guarantee the integrity and confidentiality of the
transferred messages based on Message Authentication Codes (MACs) and symmetric encryption. Today, secure communication channels are very efficient and widely
used and several standards and technologies exist, e.g., Transport Layer Security
(TLS),[1] or Internet Protocol Security.[2]

---

[1] http://datatracker.ietf.org/wg/tls/
[2] http://datatracker.ietf.org/wg/ipsecme/

T. Schneider, *Engineering Secure Two-Party Computation Protocols*,
DOI: 10.1007/978-3-642-30042-4_1, © Springer-Verlag Berlin Heidelberg 2012

Table 1.1 Comparison of PET

	Secure channels	Trusted comput- ing	Secure computa- tion
Security: Protect against malicious	Outsiders	Endpoints (partial)	Endpoints (full)
Trust: Endpoints trusted	Fully	Partially	Not
Performance	High	Medium	Low
Trusted hardware	Optional	Required	Optional
Deployment	Widely used	Standards and products	First prototypes

While secure communication channels protect against outside attackers, they assume that the communicating endpoints trust each other and potentially also a Public-Key Infrastructure (PKI) for key management.

Trusted Computing. When endpoints can no longer be trusted entirely, trusted computing technology, e.g., as standardized by the Truster Computing Group,[3] can be used to build systems that can be trusted to some extent. These trusted systems make use of a trust anchor in the form of a secure hardware module, e.g., a Trusted Platform Module (TPM) [216], and allow it to protect cryptographic keys, authenticate the configuration of a platform (attestation), and cryptographically bind sensitive data to a certain system configuration (sealing) [187].

Even though trusted computing can be used to greatly enhance trust in remote systems, there are still unsolved problems such as runtime integrity or hardware attacks on the trust anchor [114].

Secure Computation. A method that allows two or more parties to jointly perform computations without trusting each other was proposed by Yao [230]. In this book we concentrate on the two-party client/server setting, called Secure Function Evaluation (SFE) , where two mutually mistrusting parties compute an arbitrary function f on their private inputs x, y without the help of a Trusted Third Party (TTP) while revealing no information about their inputs beyond the result $f(x, y)$.

Using Garbled Circuits (GCs) [231], an arbitrary function can be computed securely in a constant number of rounds with computation and communication linear in the size of the function. The basic idea is that one party creates the GC by "encrypting" the circuit (using symmetric keys), the other party obliviously obtains the keys corresponding to both parties' inputs and the GC, and is able to decrypt the corresponding output.

An alternative approach is based on Homomorphic Encryption (HE). Here, one party sends its encrypted inputs to the other party, who then computes the desired function under encryption using the homomorphic properties of the cryptosystem. In the end, the encrypted result is sent back and decrypted. Popular homomorphic

---

[3] http://www.trustedcomputinggroup.org

cryptosystems are the additively homomorphic cryptosystems of Paillier [176] and Damgård-Jurik [65] which require interaction for multiplication. The recently proposed fully homomorphic schemes [92, 93, 201, 223] allow both addition and multiplication under encryption, but still need to be improved to become usable in practical applications.

For many years, these two approaches for SFE, GC and HE, have co-existed with their respective advantages and drawbacks: e.g., GC is highly efficient as it is mostly based on symmetric key cryptography, but requires communication proportional to the size of the evaluated function; HE can result in less communication overhead, but requires expensive public-key operations.

In recent years, SFE was used as an enabling technology for a large number of security- and privacy-critical applications (e.g., electronic auctions [164], data mining [152], remote diagnostics [47], medical diagnostics [19], or face recognition [76]. To bring such protocols closer to deployment in real-world applications, we apply several ideas from Algorithm Engineering to SFE as described next.

1.2 Outline

Algorithm Engineering (AE). "AE is concerned with the design, analysis, implementation, tuning, debugging and experimental evaluation of computer programs for solving algorithmic problems. It provides methodologies and tools for developing and engineering efficient algorithmic codes and aims at integrating and reinforcing traditional theoretical approaches for the design and analysis of algorithms and data structures" [71].

The major goals of AE are to bridge the gap between theory and practice, accelerate the transfer of algorithmic results into applications, and keep the advantages of theoretical treatment such as the generality of the solutions. In particular, AE tries to find algorithms that work well in *practice*, i.e., are simple, re-usable, and also take constant factors into account [191].

In this book we apply AE to SFE in order to engineer efficient SFE protocols for several practical applications. We focus on the following aspects presented in the respective chapters:

Basic Techniques (Sect. 2). We start with a detailed review of state-of-the-art techniques and optimizations for SFE with special focus on their practicability. In particular we are concerned with their performance including constant factors and possibilities for pre-computation. For GC-based SFE protocols we show how most of the complexity can be shifted into a setup phase.

Efficiency (Sect. 3). As the efficiency of SFE protocols depends on the size of the evaluated function, a small representation of the function should be chosen. We show how to reduce the function size and provide efficient building blocks for standard functionalities.

Deployability (Sect. 4). When SFE protocols are deployed in real-world systems, available Hardware (HW) resources can be used to improve the SFE protocols.

Besides using the available HW as an accelerator for more efficient computation, we also show how secure HW can be used to make the communication of GC-based SFE protocols *independent* of the size of the evaluated function.

Usability (Sect. 5). In order to make cryptographic protocols usable also for non-experts, compilers have been developed that automatically generate protocols from high-level descriptions, e.g., within the CACE project [14]. Examples of protocols that can be generated automatically include Zero-Knowledge Proofs of Knowledge [7, 15–18, 48, 54, 158], and Secure Computation [25, 32, 69, 109, 143, 155, 157, 172, 177, 195, 196, 199]. While previous compilers for secure computation have been restricted to use either HE or GCs, we present a framework to combine both approaches in a modular and secure way, and provide a novel tool and language for the specification and automatic generation of such hybrid SFE protocols, called Tool for Automating Secure Two-party Computations (TASTY).

Conclusion (Sect. 6). Finally, we summarize the contents of this book by giving guidelines for designing efficient SFE protocols and directions for future work.

Chapter 2
Basics of Efficient Secure Function Evaluation

2.1 Common Notation and Definitions

In this section we introduce common notation (Sect. 2.1.1), cryptographic primitives (Sect. 2.1.2), function representations (Sect. 2.1.3), the adversary model (Sect. 2.1.4), and the Random Oracle (RO) model (Sect. 2.1.5) used in this book.

2.1.1 Notation

We use the following standard notations.

2.1.1.1 Basics

Bitstrings. $\{0, 1\}^\ell$ denotes the space of binary strings of length ℓ. $a||b$ denotes the concatenation of strings a and b. $\langle a, b \rangle$ is a vector with two components a and b, and its representation as a bit string is $a||b$. For strings $s, t \in \{0, 1\}^\ell$, $s \oplus t$ denotes their *bitwise exclusive-or* (XOR).

Random Choice. Uniform random choice is denoted by the $\in_R$ operator, e.g., $r \in_R D$ reads "draw r uniformly at random from D".

Protocol Participants. We call the two Secure Function Evaluation (SFE) participants *client* C (Alice) and *server* S (Bob). This naming choice is influenced by the asymmetry in the SFE protocols, which fits into the client–server model. We want to point out that we do not limit ourselves to this setting even though this client–server relationship in fact exists in most real-life two-party SFE scenarios.

2.1.1.2 Security and Correctness Parameters

Our security and correctness parameters are named as shown in Table 2.1.

Table 2.2 contains current recommendations by ECRYPT II [74] for the size of the symmetric security parameter t and the asymmetric security parameter T.

T. Schneider, *Engineering Secure Two-Party Computation Protocols*, 5
DOI: 10.1007/978-3-642-30042-4_2, © Springer-Verlag Berlin Heidelberg 2012

Table 2.1 Security and correctness parameters

Symbol	Name
t	Symmetric security parameter (bit length of symmetric keys)
T	Asymmetric security parameter (bit length of RSA moduli)
σ	Statistical security parameter
κ	Correctness parameter

Table 2.2 Security parameters: recommended sizes [74]

Security level	Recommended use until	t (bit)	T (bit)
Ultra-short	2012	80	1,248
Short	2020	96	1,776
Medium	2030	112	2,432
Long	2040	128	3,248

An overview and comparison of different recommendations is available at [96].

In implementations, the statistical security parameter σ and the correctness parameter κ can be chosen as $\sigma = \kappa = 40$.

2.1.2 Cryptographic Primitives

Pseudo-Random Function (PRF) keyed with k and evaluated on x is denoted by $\mathsf{PRF}_k(x)$. PRF can be instantiated with a block cipher, e.g., AES, or a cryptographic hash function, e.g., SHA-256. AES is preferable if PRF is run repeatedly with the same key k as in this case the key schedule of AES needs to be run only once and hence amortizes.

Message Authentication Code (MAC) keyed with k and evaluated on message m is denoted by $\mathsf{MAC}_k(m)$. In our token-based protocols in Chap. 4 we use a MAC algorithm that does not need to store the entire message, but can operate "online" on small blocks, e.g., AES-CMAC [204] or HMAC [146].

2.1.3 Function Representations

We use several standard representations for functions which are particularly useful for SFE protocols as shown in Fig. 2.1: boolean circuits (Sect. 2.1.3.1) and arithmetic circuits (Sect. 2.1.3.2).

Fig. 2.1 Function representations. **a** Boolean circuit. **b** Arithmetic circuit

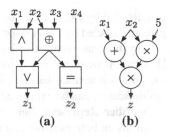

(a)   (b)

2.1.3.1 Boolean Circuits

Boolean circuits are a classical representation of functions in engineering and computer science.

Definition 1 (*Boolean Circuit*) A *boolean circuit* with u inputs, v outputs and n gates is a Directed Acyclic Graph (DAG) with $|V| = u + v + n$ vertices (nodes) and $|E|$ edges. Each node corresponds to either a *gate*, an *input* or an *output*. The edges are called *wires*.

For simplicity, the input and output nodes are often omitted in the graphical representation of a boolean circuit as shown in Fig. 2.1a. For a more detailed definition see [225].

Definition 2 (*Gate*) A d-input gate G_d is a boolean function which maps $d \geq 0$ input bits to one output bit, i.e., $G_d : (in_1, \ldots, in_d) \in \{0, 1\}^d \to \{0, 1\}$.

Typical gates are XOR ($\oplus$), XNOR ($=$), AND ($\wedge$), OR ($\vee$).

Topologic Order. Gates of a boolean circuit can be evaluated in any order, as long as all of the current gate's inputs are known. This property is ensured by sorting (and evaluating) the gates topologically, which can be done efficiently in $O(|V| + |E|)$ [64, Topological sort, pp. 549–552]. The topologic order of a boolean circuit indexes the gates with labels $G_1, \ldots, G_n$ and ensures that the ith gate G_i has no inputs that are outputs of a successive gate $G_{j>i}$. In complexity theory, a circuit with such a topologic order is called a *straight-line program* [6]. Given the values of the inputs, the output of the boolean circuit can be evaluated by evaluating the gates one-by-one in topologic order. A valid topologic order for the example boolean circuit in Fig. 2.1a would be $\wedge$, $\oplus$, $\vee$, $=$. The topologic order is not necessarily unique, e.g., $\oplus$, $\wedge$, $=$, $\vee$ would be possible as well.

Throughout this book we assume that boolean circuits are ordered topologically.

2.1.3.2 Arithmetic Circuits

Arithmetic circuits are a more compact function representation than boolean circuits.

An *arithmetic circuit* over a ring R and the set of variables $x_1, \ldots, x_n$ is a DAG. Figure 2.1a illustrates an example. Each node with in-degree zero is called an input gate labeled by either a variable x_i or an element in R. Every other node is called a gate and labeled by either $+$ or $\times$ denoting addition or multiplication in R.

Any boolean circuit can be expressed as an arithmetic circuit over $R = \mathbb{Z}_2$. However, if we use $R = \mathbb{Z}_m$ for sufficiently large modulus m, the arithmetic circuit can be much smaller than its corresponding boolean circuit, as integer addition and multiplication can be expressed as single operations in $\mathbb{Z}_m$.

Number Representation. We note that arithmetic circuits can simulate computations on both positive and negative integers x by mapping them into elements of $\mathbb{Z}_m : x \mapsto x \bmod m$. As with all fixed precision arithmetics, over- and underflows must be avoided.

2.1.4 Adversary Model

The standard approach for formalizing and proving security of cryptographic protocols is to consider adversaries with different capabilities. In the following we give intuition for the capabilities of *malicious, covert,* and *semi-honest* adversaries. We refer to [99] for formal definitions and to [145, 152] for more detailed discussions.

Malicious adversaries, also called *active adversaries*, are the strongest type of adversaries and are allowed to arbitrarily deviate from the protocol, aiming to learn private inputs of the other parties and/or to influence the outcome of the computation. Not surprisingly, protection against such attacks is relatively expensive, as discussed in Sect. 2.3.1.2.

Covert adversaries are similar to malicious adversaries, but with the restriction that they must avoid being caught cheating. That is, a protocol in which an active attacker may gain advantage may still be considered secure if attacks are discovered with certain fixed probability (e.g., $1/2$). It is reasonable to assume that in many social, political and business scenarios the consequences of being caught overweight the gain from cheating. At the same time, protocols secure against covert adversaries can be substantially more efficient than those secure against malicious players, as shown in Sect. 2.3.1.2.

Semi-honest adversaries, also called *passive adversaries*, do not deviate from the protocol but try to infer additional information from the transcript of messages seen in the protocol. Far from trivial, this model covers many typical practical settings such as protection against insider attacks. Further, designing and evaluating the performance of protocols in the semi-honest model is a first step towards protocols with stronger security guarantees (cf. Sect. 2.3.1.2). Indeed, most protocols and implementations of protocols for practical privacy-preserving applications focus on the semi-honest model [19, 76, 164, 173, 189].

2.1.5 Random Oracle Model

Some of our constructions in this book make use of ROs [24], a relatively strong assumption. In fact, it has been shown in [57] that some (contrived) uses of RO cannot be securely instantiated with *any* hash function. Therefore, proofs in the RO model cannot be seen as proofs in the strictest mathematical sense. However, we believe that modeling cryptographic hash functions, such as SHA-256, as RO is well-justified in many practical settings because of the following reasons:

Firstly, to date, no attacks exploiting the RO assumption are known on practical systems. This holds true even in the academic context: Important attacks on SHA-1 [226] that exploit the structure of the functions were far from being practical, and simply accelerated migration to stronger primitives, which are believed secure today. While some attacks, such as the attack on MD5 [209], are in fact practical, the use of MD5 had long been considered unsafe, and [209] broke poorly managed systems. Thus we do not consider [209] an attack on properly implemented protocols. In fact, [209] and the preliminary work that led to it only support the historic fact that users of hash functions do receive weakness warnings years ahead of possible real breaks.

Further, even in well-understood and deployed real-life systems, the crypto core (which includes the employed hash functions) is almost never targeted for attacks, in favor of *much* easier to exploit implementation flaws.

In sum, we believe that making the RO assumption on the employed hash function is practically justified in our and many other settings.

2.2 Cryptographic Primitives for Secure Two-Party Computation

In this section we describe essential building blocks used in SFE protocols: Homomorphic Encryption (HE) in Sect. 2.2.1, Garbled Circuits (GCs) in Sect. 2.2.2, and Oblivious Transfer (OT) in Sect. 2.2.3.

2.2.1 Homomorphic Encryption

HE schemes are semantically secure encryption schemes which allow computation of specific operations on ciphertexts and hence can be used for secure evaluation of arithmetic circuits as described next.

Let (Gen, Enc, Dec) be a semantically secure encryption scheme with plaintext space P, ciphertext space C, and algorithms for key generation Gen, encryption Enc and decryption Dec. We write $[\![m]\!]$ for $Enc(m, r)$, where r is random.

Table 2.3 Additively homomorphic encryption schemes

Scheme	P	Ciphertext size	$\mathrm{Enc}(m, r)$
Paillier [176]	$\mathbb{Z}_N$	$2T$	$g^m r^N \bmod N^2$
Damgård–Jurik [65]	$\mathbb{Z}_{N^s}$	$(s+1)T$	$g^m r^{N^s} \bmod N^{s+1}$
Damgård–Geisler–Krøigaard [66]	$\mathbb{Z}_u$	T	$g^m h^r \bmod N$
Lifted EC-ElGamal [75]	$\mathbb{Z}_p$	$4t+2$	$(g^r, g^m h^r)$

N RSA modulus; $s \geq 1$; u small prime; p $2t$-bit prime; g, h generators

2.2.1.1 Additively Homomorphic Encryption

An *additively* HE scheme allows addition under encryption as follows. It defines an operation $+$ on plaintexts and a corresponding operation $\boxplus$ on ciphertexts, satisfying $\forall x, y \in P : [\![x]\!] \boxplus [\![y]\!] = [\![x + y]\!]$. This naturally allows for multiplication with a plaintext constant a using repeated doubling and adding: $\forall a \in \mathbb{N}, x \in P : a[\![x]\!] = [\![ax]\!]$.

Popular instantiations for additively HE schemes are summarized in Table 2.3: The Paillier cryptosystem [176] provides a T-bit plaintext space and $2T$-bit ciphertexts, where T is the size of the RSA modulus N, and is sufficient for most applications (details see below). The Damgård–Jurik cryptosystem [65] is a generalization of the Paillier cryptosystem which provides a larger plaintext space of size sT-bit for $(s + 1)T$-bit ciphertexts for arbitrary $s \geq 1$. The Damgård–Geisler–Krøigaard cryptosystem [66–68] has smaller ciphertexts of size T-bit, but can only be used with a small plaintext space $\mathbb{Z}_u$, where u is a small prime, as decryption requires computation of a discrete logarithm. Finally, the lifted ElGamal [75] cryptosystem has additively homomorphic properties and very small ciphertexts. However, decryption is only possible if the plaintext is known to be in a small subset of the plaintext space, as the discrete logarithm of a generator with large order has to be brute-forced. Lifted ElGamal implemented over an EC group (Lifted EC-ElGamal) provides a $2t$-bit plaintext space and very small ciphertexts of size $2(2t + 1)$ bits $\sim 4t$ bits when using point compression.

The Paillier Cryptosystem. The most widely used additively HE system is that of Paillier [176] where the public key is an RSA modulus N and the secret key is the factorization of N. The extension of [65, Sect. 6] allows one to pre-compute expensive modular exponentiations of the form $r^N \bmod N^2$ in a setup phase, s.t. only two modular multiplications per encryption are needed in the time-critical online phase. The party which knows the factorization of N (i.e., the secret key), can use Chinese remaindering to efficiently pre-compute these exponentiations and to decrypt.

2.2.1.2 Fully Homomorphic Encryption

Some HE systems allow both addition and multiplication under encryption. For this, a separate operation $\times$ for multiplication of plaintexts and a corresponding operation $\boxtimes$ on ciphertexts is defined satisfying $\forall x, y \in P : [\![x]\!] \boxtimes [\![y]\!] = [\![x \times y]\!]$. Cryptosystems with such a property are called *fully* homomorphic.

Until recently, it was widely believed that such cryptosystems do not exist. Several works provided partial solutions: [34] and [95] allow for polynomially many additions and one multiplication, and ciphertexts of [193] grow exponentially in the number of multiplications. Recent schemes [92, 93, 201, 223] are fully homomorphic. Even if the size of the ciphertexts of these fully HE schemes is independent of the number of multiplications, the size and computational cost of fully HE schemes are *substantially* larger than those of additively HE schemes. First implementation results of [201] show that even for almost fully HE schemes with conservatively chosen security parameters that allow for multiplicative depth $d = 2.5$ of the evaluated circuit, i.e., at most two multiplications, encrypting a single bit takes 3.7 s on a 2.4 GHz Intel Core2 (6600) CPU. Most recent implementation results of [148] indicate that the performance of somewhat homomorphic encryption schemes might be sufficient for outsourcing certain types of computations, whereas fully HE is still very inefficient as shown in [94] whose implementation requires in the order of Gigabytes of communication and minutes of computation on high-end IBM System $\times 3500$ servers.

Although significant effort is underway in the theoretical community to improve its performance, it seems unlikely that fully HE will reach the efficiency of current public-key encryption schemes. Intuitively, this is because a fully HE cryptosystem must provide both the same strong security guarantees (semantic security) and extra algebraic structure to allow for homomorphic operations. The extra structure weakens security, and countermeasures (costing performance) are necessary.

2.2.1.3 Computing on Encrypted Data

Homomorphic encryption naturally allows for secure evaluation of arithmetic circuits over P, called computing on encrypted data, as follows. The client $\mathcal{C}$ generates a key pair for an HE cryptosystem and sends her inputs encrypted under the public key to the server $\mathcal{S}$ together with the public key. With a fully HE scheme, $\mathcal{S}$ can simply evaluate the arithmetic circuit by computing on the encrypted data and send back the (encrypted) result to $\mathcal{C}$, who then decrypts it to obtain the output.[1] If the HE scheme only supports addition, one round of interaction between $\mathcal{C}$ and $\mathcal{S}$ is needed to evaluate each multiplication gate (or layer of multiplication gates) as described below. Today,

---

[1] If $\mathcal{S}$ is malicious, it must additionally be ensured that he indeed computed the intended functionality by means of verifiable computing (cf. Sect. 4.3.3.2).

using additively HE and interaction for multiplication results in much faster SFE protocols than using fully HE schemes.

Packing. Often the plaintext space P of the HE scheme is substantially larger than the size of the encrypted numbers. This allows one to pack multiple numbers into one ciphertext (before or after additive blinding) and send back only a single ciphertext from S to C. This optimization substantially decreases the size of the messages sent from S to C as well as the number of decryptions performed by C. The computational overhead for S is small as packing the ciphertexts $[\![x_1]\!], \ldots, [\![x_n]\!]$ into one ciphertext $[\![X]\!] = [\![x_n||\ldots||x_1]\!]$ costs less than one full-range modular exponentiation by using Horner's scheme: $[\![X]\!] = [\![x_n]\!]$; for $i = n-1$ downto $1 : [\![X]\!] = 2^{|x_{i+1}|}[\![X]\!] \boxplus [\![x_i]\!]$.

Interactive Multiplication with Additively Homomorphic Encryption. To multiply two ℓ-bit values encrypted under additively HE and held by S, $[\![x]\!]$ and $[\![y]\!]$, the following standard protocol requires one single round of interaction between S and C: S randomly chooses $r_x, r_y \in_R \{0, 1\}^{\ell+\sigma}$, where σ is the statistical security parameter, computes the blinded values $[\![\bar{x}]\!] = [\![x + r_x]\!]$, $[\![\bar{y}]\!] = [\![y + r_y]\!]$ and sends these to C. C decrypts, multiplies and sends back $[\![z]\!] = [\![\bar{x}\bar{y}]\!]$. S obtains $[\![xy]\!]$ by computing $[\![xy]\!] = [\![z]\!] \boxplus (-r_x)[\![y]\!] \boxplus (-r_y)[\![x]\!] \boxplus [\![-r_x r_y]\!]$.

Efficiency of single or multiple multiplications in parallel can be improved by *packing* the blinded ciphertexts together instead of sending them to C separately.

Security. We note that in SFE protocols based on HE, the security of the party who knows the secret key (C in our setting) is *computational* as it is computationally hard for the other party to break the semantic security of the HE scheme. The security of the party who computes under HE (S in our setting) is *statistical* as this party always statistically blinds all intermediate values before sending them back.

Efficiency. As SFE based on additively HE requires interaction for multiplying two ciphertexts, the *round complexity* of such protocols is determined by the multiplicative depth of the evaluated function, i.e., the number of successive multiplications.

When the public key is known to both parties, encryptions and re-randomization values can be pre-computed in a *setup phase*.

Still, the *online phase* requires computationally expensive public key operations such as modular exponentiations for multiplying with constants, or decryptions.

2.2.2 Garbled Circuit Constructions

Another efficient method for computing under encryption is based on Garbled Circuits (GCs). The fundamental idea of GCs going back to Yao [231], is to represent the function f to be evaluated as a boolean circuit C and encrypt (*garble*) each wire with a symmetric encryption scheme. In contrast to HE (cf. Sect. 2.2.1), the encryptions here cannot be operated on directly, but require helper information which is generated and exchanged in a setup phase in the form of a *garbled table* for each gate.

In this section we summarize existing schemes for constructing and evaluating GCs and give applications in Sect. 2.3.1. We give an algorithmic description of GCs and refer to the original papers on GCs constructions for details and proofs of security.

2.2.2.1 Components of GC Constructions

We start by briefly introducing the main components of GC constructions: garbled values and garbled tables to compute thereon.

Garbled Values. Computations in a GC are not performed on plain values 0 or 1, but on random-looking secrets, called *garbled values*. During construction of the GC, two random-looking garbled values $\widetilde{w}_i^0, \widetilde{w}_i^1$ are assigned to each wire w_i of C. The garbled value $\widetilde{w}_i^j$ corresponds to the plain value j, but, as it looks random, does not reveal its corresponding plain value j.

In efficient GC constructions, each garbled value is composed of a symmetric t-bit *key* and a random-looking *permutation bit* (see Point-and-Permute below):

$$\widetilde{w}_i^0 = \left\langle k_i^0, \pi_i^0 \right\rangle, \widetilde{w}_i^1 = \left\langle k_i^1, \pi_i^1 \right\rangle \quad \text{with } k_i^0, k_i^1 \in \{0, 1\}^t, \pi_i^0 \in \{0, 1\} \quad (2.1)$$

and

$$\pi_i^1 = 1 - \pi_i^0. \quad (2.2)$$

The exact method for choosing the values k_i^0, k_i^1, π_i^0 is determined by the specific GC construction (cf. Sect. 2.2.2.3).

Garbled Tables. To allow computations on garbled values, for each gate G_i ($i = 1, \ldots, n$) of the circuit C, a garbled table $\widetilde{G}_i$ is constructed. Given the garbled values corresponding to G_i's input wires, $\widetilde{G}_i$ allows one to decrypt *only* the corresponding garbled value of G_i's output wire. Formally, let $\text{in}_1, \ldots, \text{in}_d$ be the input wires of gate G and out be its output wire. Then, for *any* input combination $b_j \in \{0, 1\}$ ($j = 1, \ldots, d$), given the corresponding garbled inputs $\widetilde{\text{in}}_1^{b_1}, \ldots, \widetilde{\text{in}}_d^{b_d}$, the garbled table $\widetilde{G}$ allows one to decrypt *only* $\widetilde{\text{out}}^{G(b_1, \ldots, b_d)}$.

In particular, no information about the other garbled output value, the plain input bits b_j, or the plain output bit $G(b_1, \ldots, b_d)$ is revealed.

The general idea for constructing garbled tables is to use for all possible input combinations b_j the garbled input keys $\widetilde{\text{in}}_j^{b_j}$ to symmetrically encrypt the corresponding output key $\widetilde{\text{out}}^{G(b_1, \ldots, b_d)}$. The entries of the garbled table are the ciphertexts for all possible input combinations. The position of the entries in the garbled table must be such that it does not reveal any information about the corresponding plain input values b_j.

To achieve this, the original GC construction proposed by Yao [231] randomly permutes the entries in the garbled table. In order to find the right entry to decrypt, the symmetric encryption function requires an *efficiently verifiable range* to determine which entry was decrypted successfully, as described in [151]. However, this method has a large overhead as multiple trial-decryption need to be performed and ciphertext size increases.

In the following we briefly discuss the state-of-the-art for efficiently instantiating the used symmetric encryption function.

Point-and-Permute. The *point-and-permute* technique described in [164] allows one to immediately find the right entry for decryption in the garbled table as follows: The entries of the garbled table are permuted such that the permutation bits of the garbled input wires $\pi_1, \ldots, \pi_d$ are used to point directly to the entry of the garbled table which needs to be decrypted. As the permutation bits look random, the position of the entries in the garbled table appears random as well and hence reveals no information about the input bits $b_1, \ldots, b_d$.

By applying the point-and-permute technique, the employed symmetric encryption scheme no longer needs to have an efficiently verifiable range.

Encryption Function. The encryption function for encrypting garbled table entries is $E_{k_1,\ldots,k_d}^s(m)$ with inputs d keys of length t, a message m, and some additional information s. The additional information s must be unique per invocation of the encryption function, i.e., it is used only once for any choice of keys. Indeed, it is crucial that in the GC constructions s contains a unique and independent gate identifier (cf. [213]).

As proposed in [154, 180], E can be instantiated efficiently with a *Key Derivation Function* $\mathsf{KDF}^\ell(k_1, \ldots, k_d, s)$ whose ℓ bits of output are independent of the input keys $k_1, \ldots, k_d$ in isolation, and which depends on the value of s:

$$E_{k_1,\ldots,k_d}^s(m) = m \oplus \mathsf{KDF}^{|m|}(k_1, \ldots, k_d, s). \tag{2.3}$$

KDF can be instantiated with a cryptographic hash function H:

- The most efficient implementation of KDF is a single invocation of H,

$$\mathsf{KDF}^\ell(k_1, \ldots, k_d, s) = H(k_1 || \ldots || k_d || s)_{1\ldots\ell}. \tag{2.4}$$

- Alternatively, KDF could also be implemented by d separate calls to H,

$$\mathsf{KDF}^\ell(k_1, \ldots, k_d, s) = H(k_1 || s)_{1\ldots\ell} \oplus \cdots \oplus H(k_d || s)_{1\ldots\ell}. \tag{2.5}$$

In practical implementations, H can be chosen for example from the SHA-family. For provable security of the GC construction, H is modeled as RO, circular correlation robust, or PRF, depending on the specific GC construction used as described later in Sect. 2.2.2.5.

Fig. 2.2 Interface of GC constructions. **a** createGC. **b** evalGC

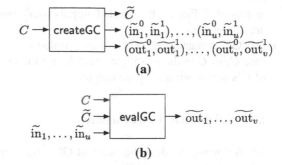

(a)

(b)

2.2.2.2 Interfaces and Structure of GC Constructions

GC constructions can be seen as algorithms with clean interfaces and a common general structure as described next.

Interface of GC Constructions. Each GC construction consists of two randomized algorithms: createGC generates a GC and evalGC evaluates it, as shown in Fig. 2.2:

- createGC takes a boolean circuit C as input and outputs the corresponding GC $\tilde{C}$ (consisting of a garbled table for each of its gates), and pairs of garbled values for each of C's input and output wires.
- evalGC gets as inputs C, $\tilde{C}$ and one garbled value for each of C's inputs, $\widetilde{in}_1, \ldots, \widetilde{in}_u$ and returns the corresponding garbled output values $\widetilde{out}_1, \ldots, \widetilde{out}_v$.

We note that the inputs and outputs of both algorithms can be streams of data, i.e., given piece-by-piece without ever storing the entire objects.

Completeness and Correctness. Each GC construction must be complete and correct. Completeness requires that for all boolean circuits C, createGC creates a GC $\tilde{C}$, pairs of garbled inputs and garbled outputs. Correctness requires that afterwards for all possible input bits $x_i \in \{0, 1\}, i = 1, \ldots, u$, given the corresponding garbled values $\widetilde{in}_i = \widetilde{in}_i^{x_i}$ as inputs, evalGC outputs the garbled values $\widetilde{out}_j = \widetilde{out}_j^{z_j}, j = 1, \ldots, v$ which correspond to C evaluated on the input values: $(z_1, \ldots, z_v) = C(x_1, \ldots, x_u)$.

One-time use. We stress that for security reasons, $\tilde{C}$ cannot be evaluated more than once (otherwise, multiple runs of evalGC can leak information about input or output values). evalGC must always be run on freshly generated outputs of createGC.

General Structure of GC Constructions. The efficient GC constructions presented next have the following general structure:

- createGC starts by assigning random-looking garbled values $(\widetilde{in}_i^0, \widetilde{in}_i^1)$ to all input wires of C and outputs these. Afterwards, for each gate G_i of C in topologic order (cf. assumption in Sect. 2.1.3.1), two random-looking garbled values are assigned to the gate's output wire and afterwards its garbled table $\tilde{G}_i$ is created and output as part of $\widetilde{C}$. Finally, the garbled outputs $(\widetilde{\text{out}}_j^0, \widetilde{\text{out}}_j^1)$ for each of C's output wires are output.

- **evalGC** evaluates the GC $\widetilde{C}$ on the garbled inputs $\widetilde{\text{in}}_i$ by evaluating each garbled gate $\widetilde{G}_i$ of $\widetilde{C}$ in the topologic order determined by C. Finally, the garbled values of C's output wires $\widetilde{\text{out}}_j$ are output.

2.2.2.3 Efficient GC Constructions

In the following we describe efficient GC constructions which are suited well for efficient implementation [180].

All GC constructions presented next start with choosing the garbled input values. The garbled zero values $\widetilde{\text{in}}_i^0$ are chosen randomly, i.e., $k_i^0 \in_R \{0, 1\}^t$ and $\pi_i^0 \in_R \{0, 1\}$ in Eq. (2.1). The corresponding garbled values for one $\widetilde{\text{in}}_i^1$ are chosen randomly as $k_i^1 \in_R \{0, 1\}^t$, or according to Eq. (2.8) in the case of "free XOR" as described below.

The following GC techniques successively fix the garbled output values of each gate in order to decrease the size of the garbled tables.

Point-and-Permute. The point-and-permute GC construction was first described in [164], implemented in Fairplay [157], and also used in [142, 154].[2] This technique chooses both garbled output values of a d-input gate G_i at random and results in a garbled table with 2^d table entries. For each of the 2^d possible input combinations $b_1, \ldots, b_d$, the garbled table entry at position $\pi_1, \ldots, \pi_d$ is constructed by using the keys of G_i's garbled inputs to encrypt the corresponding garbled output:

$$\pi_1, \ldots, \pi_d : E_{k_1^{b_1}, \ldots, k_d^{b_d}}^{i \| \pi_1 \| \ldots \| \pi_d} \left(\widetilde{\text{out}}^{G(b_1, \ldots, b_d)} \right). \tag{2.6}$$

Garbled Row Reduction. The GC construction of [164], called *Garbled Row Reduction*, extends the point-and-permute GC construction by fixing one of the garbled output values resulting in a garbled table of $2^d - 1$ table entries. The first entry of each garbled table is forced to be zero and hence does not need to be transferred. By substituting into Eq. (2.6), this fixes one of the two garbled output values to be pseudo-randomly derived from the garbled input values. The other garbled output value is chosen at random satisfying Eq. (2.2). For details we refer to the description in [180].

Secret-Sharing. The GC construction of Pinkas et al. [180] uses Shamir's secret-sharing [197] to fix both garbled output values resulting in a garbled table with $2^d - 2$ entries. In the following we summarize the general idea of this construction and refer to [180, Sect. 5] for details.

The construction exploits the fact that both keys of a gate's garbled output values can be chosen independently and pseudo-randomly. The basic idea is to

---

[2] We note that the GC construction of Yu et al. [233, Sect. 3.3] is less efficient as garbled tables are larger and require slightly more computation.

pseudo-randomly derive keys $K_r \in \{0, 1\}^t$ and bit masks $M_r \in \{0, 1\}$ for all combinations of garbled inputs as

$$K_r \| M_r = \mathsf{KDF}^{t+1} \left(k_1^{b_1} \| \ldots \| k_d^{b_d} \| s \right). \tag{2.7}$$

The keys K_r are interpreted as elements in $\mathbb{F}_{2^t}$ and used as supporting points of two polynomials $P(X)$, $Q(X)$ of the same degree: $P(X)$ is defined by those keys which should map to the garbled output value $\widetilde{out}^1 := P(0)$. Similarly, $Q(X)$ maps to the garbled output value $\widetilde{out}^0 := Q(0)$. Overall, $2^d - 2$ points are stored as part of the garbled table, where some points are on both and some on only one of the polynomials. The bits M_r are used to encrypt the permutation bits of the garbled outputs as in the point-and-permute GC construction resulting in an additional 2^d encrypted bits in the garbled table.

During evaluation of the garbled gate, the garbled inputs are used to derive K_r, M_r according to Eq. (2.7). Then, M_r is used to decrypt the output permutation bit which defines through which of the supporting points in the garbled table to interpolate the polynomial. Finally, the garbled output key is determined by evaluating the polynomial at $X = 0$.

Generalization to arbitrary d. We note that the Secret-Sharing GC construction can be generalized from $d = 2$ (as described in [180, Sect. 5]) to arbitrary d-input gates as follows: Assume that n_1 of the 2^d entries in the gate's function table equal one and the remaining $n_0 := 2^d - n_1$ entries equal zero. In the following we assume that $n_1 \geq n_0$ (otherwise we invert the role of zero and one). The polynomial P, interpolated through those keys K_r that should map to the garbled output value for one, has degree n_1. We store $n_1 - 1$ extra points $P(2^d + 1), \ldots, P(2^d + n_1 - 1)$ in the garbled table. Afterwards, we interpolate polynomial Q of degree n_1 through the n_0 keys K_r that should map to the garbled output value for zero and the common $n_1 - n_0$ extra points $P(2^d + 1), \ldots, P(2^d + n_1 - n_0)$. Now, we create $n_0 - 1$ extra points $Q(2^d + n_1 - n_0 + 1), \ldots, Q(2^d + n_1 - 1)$. The order of the extra points on P and Q in the garbled table is such that the output permutation bit can be used to obliviously index which extra points to use for interpolation. The garbled table consists of $n_1 - n_0$ common extra points and $n_0 - 1$ extra points on P resp. Q, in total $n_1 - n_0 + 2(n_0 - 1) = n_1 + n_0 - 2 = 2^d - 2$ keys. The overall size of the garbled table hence is $(2^d - 2)t + 2^d$ bits.

Free XOR. As observed in [142], a fixed distance between corresponding garbled values allows "free" evaluation of XOR gates, i.e., garbled XOR gates require no garbled table and allow very efficient creation and evaluation (XOR of the garbled values). The main idea is to choose a fixed relation between the two garbled values for each garbled wire:

$$k_i^1 := k_i^0 \oplus \Delta, \tag{2.8}$$

where $\Delta \in_R \{0, 1\}^t$ is the randomly chosen *global key distance*. During creation of a garbled XOR gate, the garbled output value is set to $\widetilde{out}^0 = \widetilde{in}_1^0 \oplus \widetilde{in}_2^0$. Similarly,

Table 2.4 Efficient GC constructions for d-input gates

GC technique	Size of garbled table (bits)	Free XOR [142]
Point-and-permute [164]	$2^d t + 2^d$	✓
Garbled row reduction [164]	$(2^d - 1)t + (2^d - 1)$	✓
Secret-sharing [180]	$(2^d - 2)t + 2^d$	✗

t: symmetric security parameter

evaluation of a garbled XOR gate is done by computing $\widetilde{\text{out}} = \widetilde{\text{in}}_1^0 \oplus \widetilde{\text{in}}_2^0$. Garbled non-XOR gates can be constructed with any GC construction which fixes at most one of the garbled outputs of a gate, i.e., from the GC techniques described above Point-and-Permute and Garbled Row Reduction allow combination with "free XORs", but not the Secret-Sharing technique (cf. Table 2.4).

2.2.2.4 Complexity of Efficient GC Constructions

The complexity of the GC constructions presented in Sect. 2.2.2.3 is summarized in Table 2.4. When using free XORs, XOR gates require no communication and only negligible computation (XOR of bitstrings). We compare the complexity for other gates next.

Computation Complexity. Interestingly, all GC constructions have almost the same computation complexity, which is dominated by invocations of a cryptographic hash function H: for each d-input gate, createGC requires 2^d invocations of KDF and evalGC requires one invocation. As described in Sect. 2.2.2.1, each invocation of KDF needs one or d invocations of H depending on whether H is modeled as RO or not.

The Secret-Sharing GC construction requires slightly more computations as it also requires interpolation of two polynomials of degree at most $2^d - 1$ over $\mathbb{F}_{2^t}$.

On the other hand, the computation complexity to randomly choose the garbled output values of the gates decreases as follows: Point-and-Permute chooses both garbled values (one with free XOR), Garbled Row Reduction one (none with free XOR), and Secret-Sharing none.

Communication Complexity. As shown in Table 2.4, the size of each garbled table decreases by approximately t bits per gate from Point-and-Permute to Garbled Row Reduction and from there to Secret-Sharing. Especially for gates with low degree d these savings can be quite significant, i.e., up to -25% for Garbled Row Reduction and -50% for Secret-Sharing for the common case of $d = 2$.

However, the Secret-Sharing construction, which cannot be combined with Free XOR, results only in better communication complexity than Garbled Row Reduction if the evaluated circuits do not have many XOR gates. Indeed, we show in Chap. 3 that most commonly used circuit building blocks can be transformed such that most of the gates are XOR gates and hence Garbled Row Reduction is more efficient than Secret-Sharing w.r.t. both computation and communication.

2.2.2.5 Security of Efficient GC Constructions

The first full proof of security of the original version of Yao's GC protocol [231] was given in [151]. This proof was later adapted to show the security of various efficient GC constructions that differ in how the underlying KDF is composed from calls to H (cf. Eqs. 2.4 vs. 2.5) and how H needs to be modeled.

For practical applications, modeling H as a RO and instantiating it with a call to a cryptographic hash function, e.g., chosen from the SHA family, should provide reasonable security guarantees for all efficient GC constructions presented above. In more detail, the current situation is as follows:

The GC construction that uses Point-and-Permute together with free XORs and instantiates KDF with a single invocation of H (cf. Eq. 2.4) was proven secure when H is modeled as RO [142]. As proven in [60], this assumption can be relaxed to circular correlation robustness, but not to correlation robustness alone.

According to [154], for Point-and-Permute without free XORs, H can be modeled as RO for one invocation of H (cf. Eq. 2.4), and as PRF for several invocations of H (cf. Eq. 2.5).

As sketched in [180], for Garbled Row Reduction and Secret-Sharing, that use several invocations of H (cf. Eq. 2.5), H can be modeled to be some variant of correlation robust or as PRF, depending on whether free XORs are used or not.

2.2.3 Oblivious Transfer

Parallel 1-out-of-2 OT of n t'-bit strings, denoted as $OT_{t'}^n$, is a two-party protocol run between a chooser (client C) and a sender (server S) as shown in Fig. 2.3: For $i = 1, \ldots, n$, S inputs n pairs of t'-bit strings $s_i^0, s_i^1 \in \{0, 1\}^{t'}$ and C inputs n choice bits $b_i \in \{0, 1\}$. At the end of the protocol, C learns the chosen strings $s_i^{b_i}$, but nothing about the other strings $s_i^{1-b_i}$, while S learns nothing about C's choices b_i.

In the following, we assume that OT is used in the context of SFE protocols (as described later in Sect. 2.3.1), i.e., the transferred strings are garbled values with length $t' = t + 1 \sim t$ where t is the symmetric security parameter (cf. Sect. 2.1.1.2).

We describe techniques to efficiently implement OT next.

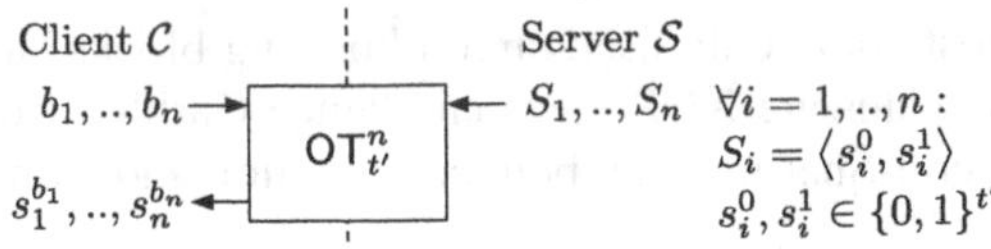

Fig. 2.3 Parallel 1-out-of-2
OT of n t'-bit strings (OT$_{t'}^n$)

Client $\mathcal{C}$ Server $\mathcal{S}$

$b_1, .., b_n \rightarrow$ ┌──────┐ $\leftarrow S_1, .., S_n$ $\forall i = 1, .., n :$
                    │ OT$_{t'}^n$ │ $S_i = \langle s_i^0, s_i^1 \rangle$
$s_1^{b_1}, .., s_n^{b_n} \leftarrow$ └──────┘ $s_i^0, s_i^1 \in \{0,1\}^{t'}$

2.2.3.1 Efficient OT Protocols

OT$_{t'}^n$ can be instantiated efficiently with different protocols, e.g., [3, 163].

For example the protocol of Naor and Pinkas [163] implemented over a suitably chosen EC consists of three messages ($\mathcal{S} \rightarrow \mathcal{C} \rightarrow \mathcal{S} \rightarrow \mathcal{C}$) in which $2n + 1$ EC points and $2nt'$ encrypted bits are sent. Using point compression, each point can be represented with $2t + 1$ bits and hence the overall communication complexity of this protocol is $(2n + 1) \cdot (2t + 1) + 2nt'$ bits $\approx 6nt$ bits. As a computation, $\mathcal{S}$ performs $2n + 1$ point multiplications and $2n$ invocations of a cryptographic hash function H, modeled as RO, and $\mathcal{C}$ performs $2n$ point multiplications and n invocations of H. This protocol is provably secure against malicious $\mathcal{C}$ and semi-honest $\mathcal{S}$ in the RO model.

Similarly, the protocol of Aiello et al. [3] implemented over a suitably chosen EC using point compression has communication complexity $n(6(2t+1))+(2t+1)$ bits $\sim 12nt$ bits and is secure against malicious $\mathcal{C}$ and semi-honest $\mathcal{S}$ in the standard model as described in [144].

2.2.3.2 Extending OT Efficiently

The extensions of Ishai et al. [121] can be used to reduce the number of computationally expensive public-key operations of OT$_{t'}^n$ to be independent of n.[3] The transformation for semi-honest $\mathcal{C}$ reduces OT$_{t'}^n$ to OT$_t^t$ (with roles of $\mathcal{C}$ and $\mathcal{S}$ swapped) and a small additional overhead: one additional message, $2n(t' + t)$ bits of additional communication, and $\mathcal{O}(n)$ invocations of a correlation robust hash function H ($2n$ for $\mathcal{S}$ and n for $\mathcal{C}$) which is substantially cheaper than $\mathcal{O}(n)$ public-key operations. A slightly less efficient OT extension for malicious $\mathcal{C}$ is given in [121] and improved in [166].

2.2.3.3 Pre-Computing OT

All computationally expensive operations for OT can be shifted into a setup phase by pre-computing OT as described in Beaver [23]: In the setup phase, the parallel OT protocol is run on randomly chosen values $r_i \in_R \{0, 1\}$ by $\mathcal{C}$ and $m_i^j \in \{0, 1\}^{t'}$ by $\mathcal{S}$. In the online phase, $\mathcal{C}$ uses her random bits r_i to mask her private inputs b_i,

---

[3] This is the reason for our choice of notation OT$_{t'}^n$ instead of $n \times$ OT$^{t'}$.

Table 2.5 Complexity of OT_t^n in the RO model

Complexity			Setup phase	Online phase
For $n \leq t$: Beaver [23] + Naor and Pinkas [163]				
Communication		Moves	3	2
		Data [bits]	$6nt$	$2nt$
Computation	Client C	H	n	
		EC mult	$2n$	
	Server S	H	$2n$	
		EC mult	$2n + 1$	
For $n > t$: Beaver [23] + Ishai et al. [121] + Naor and Pinkas [163]				
Communication		Moves	4	2
		Data [bits]	$4nt + 6t^2$	$2nt$
Computation	Client C	H	$n + 2t$	
		EC mult	$2t + 1$	
	Server S	H	$2n + t$	
		EC mult	$2t$	

and sends the masked bits to S. S replies with encryptions of his private inputs s_i^j using his random masks m_i^j from the setup phase. Which input of S is masked with which random value is determined by C's message. Finally, C applies the masks m_i she received from the OT protocol in the setup phase to decrypt the correct output values $s_i^{b_i}$.

More precisely, the *setup phase* works as follows: For $i = 1, \ldots, n$, C chooses random bits $r_i \in_R \{0, 1\}$ and S chooses random masks $m_i^0, m_i^1 \in_R \{0, 1\}^{t'}$. Both parties run an $OT_{t'}^n$ protocol on these randomly chosen values, where S inputs the pairs $\langle m_i^0, m_i^1 \rangle$ and C inputs r_i and obtains the masks $m_i = m_i^{r_i}$ as output. In the *online phase*, for each $i = 1, \ldots, n$, C masks its input bits b_i with r_i as $\bar{b}_i = b_i \oplus r_i$ and sends these masked bits to S. S responds with the masked pair of t'-bit strings $\langle \bar{s}_i^0, \bar{s}_i^1 \rangle = \langle m_i^0 \oplus s_i^0, m_i^1 \oplus s_i^1 \rangle$ if $\bar{b}_i = 0$ or $\langle \bar{s}_i^0, \bar{s}_i^1 \rangle = \langle m_i^0 \oplus s_i^1, m_i^1 \oplus s_i^0 \rangle$ otherwise. C obtains $\langle \bar{s}_i^0, \bar{s}_i^1 \rangle$ and decrypts $s_i^{b_i} = \bar{s}_i^{r_i} \oplus m_i$. Overall, the online phase consists of two messages of size n bits and $2nt'$ bits and negligible computation (XOR of bitstrings).

2.2.3.4 OT Complexity

Combining the previously described improvements for pre-computing and extending OT with the efficient OT protocol of Naor and Pinkas [163] yields a highly efficient implementation of OT_t^n in the RO model as summarized in Table 2.5. Similarly, an efficient implementation in the standard model using correlation robust hashing can be obtained by combining with the protocol of Aiello et al. [3] instead.

2.3 Garbled Circuit Protocols

In this section we show how GCs are used in several protocols for secure computation in the two-party (Sect. 2.3.1) and multi-party (Sect. 2.3.2) settings. Further applications of GC such as OTP (Sect. 4.2) or verifiable computing (Sect. 4.3) are described later in this book.

2.3.1 Two-Party Secure Function Evaluation

SFE allows two parties to implement a joint computation without using a TTP. One classical example is the Millionaires Problem [231] where two millionaires want to know who is richer, without either of them revealing their net worth to the other or a TTP.

More formally, SFE is a cryptographic protocol that allows two players, client C with private input in_C and server S with private input in_S, to evaluate a function f on their private inputs:

$$(out_C, out_S) = f(in_C, in_S). \tag{2.9}$$

The SFE protocol ensures that both parties learn only their respective output, i.e., C learns out_C and S learns out_S, but nothing else about the other party's private input. In SFE, the function f is known to both parties.[4]

Intuitively, according to the real/ideal world paradigm (e.g., [55]), an SFE protocol executed in the real world is secure if and only if an adversary with defined capabilities can do no more harm to the protocol executed in the real world than in an ideal world where each party submits its input to a TTP which computes the results according to Eq. (2.9) and returns them to the respective party.

In Sect. 2.3.1.1 we start with the description of the classical SFE protocol of Yao [231] which is secure against semi-honest adversaries and summarize how this protocol can be secured against more powerful covert and malicious adversaries in Sect. 2.3.1.2. Afterwards, we show how the evaluated function itself can be hidden in Sect. 2.3.1.3.

2.3.1.1 SFE with Semi-Honest Adversaries (Yao's Protocol)

Yao's protocol [145, 151, 231] for SFE of a function f represented as a boolean circuit (cf. Sect. 2.1.3.1) works as follows:

---

[4] If needed, SFE can be extended s.t. the function is known to only one of the parties and hidden from the other as described in Sect. 2.3.1.3.

1. *Create GC:* In the *setup phase*, the *constructor* (server S) generates a GC $\tilde{f}$ using algorithm createGC as described in Sect. 2.2.2 and sends $\tilde{f}$ to the *evaluator* (client C).
2. *Encrypt Inputs:* Afterwards, in the *online phase*, the inputs of the two parties in_C, in_S are converted into the corresponding garbled input $\widetilde{in} = \{\widetilde{in}_C, \widetilde{in}_S\}$ provided to C: For S's inputs in_S, S simply sends the garbled values corresponding to his inputs to C, i.e., $\widetilde{in}_{S,i} = \widetilde{in}_{S,i}^{in_{S,i}}$. Similarly, C must obtain the garbled values $\widetilde{in}_{C,i}$ corresponding to her inputs $in_{C,i}$, but without S learning $in_{C,i}$. This can be achieved by running (in parallel for each bit $in_{C,i}$ of in_C) a 1-out-of-2 OT protocol as described in Sect. 2.2.3.
3. *Evaluate Function Under Encryption:* Now, C can evaluate the GC $\tilde{f}$ on the garbled inputs $\widetilde{in}$ using algorithm evalGC as described in Sect. 2.2.2 and obtains the garbled outputs $\widetilde{out} = \{\widetilde{out}_C, \widetilde{out}_S\}$.
4. *Decrypt Outputs:* Finally, the garbled outputs are converted into plain outputs for the respective party: For C's outputs $\widetilde{out}_C$, S reveals their permutation bits to C (this can be done already in the setup phase). For S's outputs $\widetilde{out}_S$, C sends the obtained permutation bits to S.

Security. As proven in detail in [152], Yao's protocol is secure against semi-honest adversaries.

We observe that in Yao's protocol the security of GC constructor S is *computational* as GC evaluator C can break the GC by guessing garbled input values, verify if they decrypt correctly and match them with the garbled inputs provided by S. When instantiating OT with a protocol which provides statistical security for receiver C (e.g., using the OT protocol of Naor and Pinkas [163]), the security of GC evaluator C is *statistical*.

Efficiency. The efficiency of Yao's protocol is dominated by the efficiency of the GC construction and OT for each input bit of C.

As described in Sect. 2.2.3, OT requires only a constant number of public-key operations and allows one to shift most communication and computation into the setup phase. The resulting setup phase requires one to pre-compute $|in_C|$ OTs (cf. Sect. 2.2.3.4), create the GC $\tilde{f}$ (cf. Sect. 2.2.2.4), and transfer $\tilde{f}$ to C (cf. Table 2.4).

The online phase is highly efficient as it requires only symmetric-key operations for evaluating $\tilde{f}$ (cf. Sect. 2.2.2.4), and three moves (two for the online phase of pre-computed OT and one for sending the output to S) with about $t(2|in_C| + |in_S|) + |out_S|$ bits of communication in total.

2.3.1.2 SFE with Stronger Adversaries

GC-based SFE protocols can easily be protected against a covert or malicious client C by using an OT protocol with corresponding security properties.

Efficient SFE protocols based on GC which additionally protect against a covert [12, 103] or malicious [150] server S rely on the following cut-and-choose technique: S creates multiple GCs, deterministically derived from random seeds s_i, and commits to each, e.g., by sending $\widetilde{f}_i$ or $\mathsf{Hash}(\widetilde{f}_i)$ to $\mathcal{C}$. In the covert case, $\mathcal{C}$ asks $\mathcal{S}$ to open all but one GC $\widetilde{f}_l$ by revealing the corresponding seeds $s_{i \neq l}$. For all opened functions, $\mathcal{C}$ computes $\widetilde{f}_i$ and checks that they match the commitments. The malicious case is similar, but $\mathcal{C}$ asks $\mathcal{S}$ to open half of the functions, evaluates the remaining ones and chooses the majority of their results. Additionally, it must be guaranteed that $\mathcal{S}$'s input into OT is consistent with the GCs as pointed out in [138], e.g., using committed or committing OT. The most recent construction of [153] improves over previous protocols (smaller number of GCs, completely removing the commitments, and also removing the need to increase the size of the inputs) by using a new primitive called cut-and-choose OT, an extension of parallel 1-out-of-2 OT with a cut-and-choose functionality.

The practical performance of cut-and-choose-based GC protocols has been investigated experimentally in [154, 180]: Secure evaluation of the AES functionality (a boolean circuit with 33,880 gates) between two Intel Core 2 Duos running at 3.0 GHz, with 4 GB of RAM connected by a Gigabit ethernet takes approximately 0.5 MB data transfer and 7 s for semi-honest, 8.7 MB/1 min for covert, and 400 MB/19 min for malicious adversaries [180]. This shows that protecting GC protocols against stronger adversaries comes at a relatively high prize.

For completeness, note that cut-and-choose may be avoided with SFE schemes such as [125] which prove in zero-knowledge that the GC was computed correctly and the inputs are consistent with committed inputs [88]. However, their elementary steps involve public-key operations. As estimated in [180], such protocols which apply public-key operations per gate [125, 168] often require substantially more computation than cut-and-choose-based protocols.

We further note that there are yet other approaches to malicious security such as the approach of [123] which achieves malicious security by simulating a SMPC protocol inside a secure two-party computation protocol with semi-honest security. Their precise performance comparison is a desirable but complicated undertaking, since there are several performance measures, and some schemes may work well only for certain classes of functions.

2.3.1.3 SFE with Private Functions

In some application scenarios of SFE, the evaluated function itself needs to be hidden, e.g., as it represents intellectual property of a service provider. This can be achieved by securely evaluating a Universal Circuit (UCi) which can be programmed to simulate any circuit C and hence entirely hides C (besides an upper bound on the number of inputs, number of gates and number of outputs). Efficient UCi constructions to simulate circuits consisting of up to k two-input gates are given in [143, 221]. Generalized UCis of [184] can simulate circuits consisting of d-input gates. Which UCi construction is favorable depends on the size of the simulated functionality: Small circuits can be simulated with the UCi construction of [184, 194] with overhead $\mathcal{O}(k^2)$ gates, medium-size circuits benefit from the construction of [143] with overhead $\mathcal{O}(k \log^2 k)$ gates and for very large circuits the construction of [221] with overhead $\mathcal{O}(k \log k)$ gates is most efficient. Explicit sizes and a detailed analysis of the break-even points between these constructions are given in [184]. The alternative approach of [136] for evaluating private functions without using UCis has complexity linear in k, but requires $\mathcal{O}(k)$ public-key operations.

While UCis entirely hide the structure of the evaluated functionality f, it is sometimes sufficient to hide f only within a class of topologically equivalent functionalities $\mathcal{F}$, called secure evaluation of a *semi-private* function $f \in \mathcal{F}$ [177]. The circuits for many standard functionalities are topologically equivalent and differ only in the specific function tables, e.g., comparison $(<, >, =, \dots)$ or addition/subtraction, as described later in Sect. 3.3. When no cut-and-choose is used for GCs, it is possible to directly evaluate the circuit and avoid the overhead of a UCi for semi-private functions, as GC constructions of [157] and [164] (cf. Sect. 2.2.2.3) completely hide the type of the gates from the GC evaluator. These techniques were used for example in [83–86, 177].

2.3.2 Garbled Circuit Protocols with Multiple Parties

GCs can also be used for SMPC, i.e., secure computation with more than two parties. In the following we describe applications of GCs to SMPC in Sect. 2.3.2.1 and secure mobile agents in Sect. 2.3.2.2.

In the multi-party setting, one party, the GC *creator*, which is assumed to behave correctly, creates the GC (cf. algorithm createGC in Sect. 2.2.2.2); another party, the GC *evaluator*, obliviously obtains the corresponding garbled inputs and evaluates the GC (cf. algorithm evalGC in Sect. 2.2.2.2). The other parties provide inputs to or obtain outputs from the protocol.

We will show later in Chap. 4 that the GC creator can be implemented with constant-size memory, e.g., within a tamper-proof HW token.

Verifiability of GC. As discussed in detail in Chap. 4, the GC evaluator, who evaluates the GC on the garbled inputs, need not be trusted at all. Indeed, GC evaluation can be performed by one or more *untrusted* parties as the garbled outputs allow verification that the GC evaluation was done correctly [164]: For each garbled output $\tilde{z}_i$, the GC creator provides the *output decryption information* $\langle 0, G(\tilde{z}_i^0) \rangle, \langle 1, G(\tilde{z}_i^1) \rangle$, where G is a one-way function (e.g., a cryptographic hash function). This allows one to check whether $\tilde{z}_i$ is correct, i.e., either $\tilde{z}_i = \tilde{z}_i^0$ or $\tilde{z}_i = \tilde{z}_i^1$, and which is the corresponding plain value without revealing the values $\tilde{z}_i^0$ and $\tilde{z}_i^1$. As the GC evaluator is unable to guess a correct $\tilde{z}_i$ (except with negligible probability), she must have obtained it by honestly evaluating the GC.

2.3.2.1 SMPC with Two Servers

As proposed in [164], Yao's GC protocol (cf. Sect. 2.3.1.1) can be turned into a SMPC protocol with multiple input players, multiple output players, and two non-colluding computation players who perform the secure computation: the GC creator is trusted by the output players to behave semi-honestly and the GC evaluator can even be malicious.

For multiple input players, the parallel 1-out-of-2 OT protocol (cf. Sect. 2.2.3) is replaced with a parallel 1-out-of-2 proxy OT protocol. The proxy OT protocol splits the role of the chooser in the OT protocol into two parties: the *chooser* (input player) provides the secret input bit b, and the *proxy* (the GC evaluator) learns the chosen output string s^b, but neither b nor s^{1-b}. As described in [164, Appendix A], efficient OT protocols (e.g., the protocols of Aiello et al. [3], Naor and Pinkas [163] described in Sect. 2.2.3) can be naturally converted into a proxy OT protocol as follows: The chooser sends the two public keys, of which she knows the trapdoor to exactly one, to the sender. The sender applies an error-correcting code to each of the two strings s^0, s^1 and sends their encryptions under the respective public key to the proxy. The proxy uses the trapdoor obtained by the chooser to decrypt both ciphertexts obtained from the sender and uses the error correcting code to compute s^b.

For multiple output players, the GC evaluator forwards the garbled outputs to the respective output player who can decrypt and verify the correctness of the output using the output decryption information obtained from the GC creator.

2.3.2.2 Secure Mobile Agents

In the *mobile agents* scenario, the *originator* creates SW agents that can perform tasks on behalf of the originator. After creating the agents for some specific purpose, the originator sends them out to visit various remote hosts, where the agents perform computations on behalf of the originator. When the agents return home, the originator retrieves the results of these computations from the agents. The utility of this paradigm is based on the ability of the originator to go offline after sending the agents out, and, ideally, no further interaction between the agent and the originator or the host should be required. A possible application would be an agent which travels through the web to select, depending on a policy of the originator, an offer for the most suitable product at the lowest price.

Secure mobile agents extend the mobile agents scenario with security features. Here, the visited hosts are not trusted by the originator and vice versa. When an agent visits a host, it carries along some state from previous computations and uses this together with input from the host to compute the new agent state possibly along with an output provided to the host. The agent state (both old and new) is "owned" by the agent, and should be protected from potentially malicious hosts, whereas the host input and output are "owned" by the host and should likewise be protected from potentially malicious agents. The code evaluated by the agent (policy) can be hidden as well by evaluating a UCi (cf. Sect. 2.3.1.3).

The concept of secure mobile agents was introduced in [192] who give partial solutions based on HE (Sect. 2.2.1). More practical constructions for secure mobile agents proposed afterwards are based on GCs: An agent can securely migrate from one host to the next by running a (slightly modified) GC-based SFE protocol (cf. Sect. 2.3.1) between the two hosts as described in [53]. To protect against malicious hosts, a TTP can be used to generate the GCs, similarly to the GC constructor in the construction of Naor et al. [164] (cf. Sect. 2.3.2.1), as proposed in [5]. The assumption of the TTP was later removed in [212] and a construction which achieves universal composability is given in [229]. Finally, non-interactive OT based on trusted HW reduces the communication overhead to the essential minimum where the agent is sent from one host to the next in a single message [106].

Chapter 3
Circuit Optimizations and Constructions

3.1 Introduction

We start with a motivation of protocols that allow "free" evaluation of XOR gates (Sect. 3.1.1). Afterwards we give related work (Sect. 3.1.2) and preliminaries and notation used in this chapter (Sect. 3.1.3).

3.1.1 Protocols with Free XOR

The following cryptographic protocols allow evaluation of XOR gates (addition modulo 2) at a much lower cost than non-XOR gates such as AND gates (multiplication modulo 2). All these applications benefit from our circuit optimizations of Sect. 3.2.2 which reduce the number of costly non-XOR gates.

Secure Function Evaluation (SFE). Gate evaluation secret sharing [140] allows information theoretically secure SFE where the size of shares remains constant for XOR gates and increases (doubles) for non-XOR gates. Also the Garbled Circuits (GC) construction of Kolesnikov and Schneider [142] provides "free", i.e., non-interactive and computationally inexpensive, garbled XOR gates (cf. Sect. 2.2.2.3).

Homomorphic Encryption (HE). The homomorphic cryptosystems of Boneh et al. [34], Gentry et al. [95] allow evaluation of 2-DNF formulas, i.e., polynomially many additions and one multiplication under encryption (cf. Sect. 2.2.1.2). When instantiated over $\mathbb{F}_2$, such schemes allow evaluation of many XOR gates but only one AND gate.

Recent fully HE systems [92, 93, 201, 208, 223] support evaluation of arbitrarily many additions and multiplications under encryption (cf. Sect. 2.2.1.2). However, addition (evaluation of an XOR gate) is more efficient than multiplication (evaluation of an AND gate) as it has less impact on the "error" of the ciphertext and hence fewer "refresh" (re-encrypt) operations are needed (cf. [92]).

T. Schneider, *Engineering Secure Two-Party Computation Protocols*,
DOI: 10.1007/978-3-642-30042-4_3, © Springer-Verlag Berlin Heidelberg 2012

Zero-Knowledge. Zero-knowledge interactive proofs (ZKIP) for SAT, i.e., to prove satisfiability of a formula represented as a boolean circuit in zero-knowledge, can be constructed based on the quadratic residuosity assumption [45]. This protocol allows non-interactive processing of NOT and XOR gates and requires one round of interaction for other gates. Extensions of these proof techniques of Boyar et al. [36, 39–42] inherit and rely on the property of cheap non-interactive processing of XOR gates.

3.1.2 Related Work

Existing frameworks for automatic generation of SFE protocols [25, 157, 234] allow one to specify the function to be computed in a function description language and automatically transform this description into a functionally equivalent boolean circuit. Alternatively, existing tools from electronic design automation such as compilers for Very High Speed Integrated Circuit Hardware Description Language (VHDL) or Verilog could be used for generating boolean circuits. The open-source Fairplay compiler suite [25, 157] performs optimizations on the high-level function description language, e.g., using techniques such as peek-hole optimization, duplicate code removal, or dead code elimination. Our optimizations are on the lower abstraction level of circuits and can also be applied to further optimize the output of circuits generated with the Fairplay compiler (e.g., as used in [180, Chap. 3] to minimize the AES circuit).

A special case of our algorithm for propagating 1-input gates into their predecessor or successor gates was mentioned for NOT gates in [233]. Further, they give explicit circuit sizes for standard functionalities such as addition, subtraction, multiplication, multiplexer, and comparison of unsigned integers when circuits are decomposed into 2-input gates. However, when XOR gates are not "for free", the case they consider, GC constructions are more efficient when not decomposing 3-input gates into more than two 2-input gates [154].

Gates with many inputs can be decomposed into smaller gates using Shannon's expansion theorem [198] which replaces a $(d + 1)$-input gate by two d-input gates and one bit-multiplexer (one 3-input gate or three 2-input gates). For decomposition into 2-input gates, the Quine–McCluskey algorithm can be extended to also handle XOR gates [219]. However, these algorithms use a monotonic cost metric for optimization which does not consider "free" XOR gates.

For optimizations in the presence of "free" XOR gates, Boyar et al. [36, 42, 43] considered multiplicative complexity[1] of symmetric functions, i.e., functions that only depend on the Hamming weight of their inputs and hence the inputs of such functions can be permuted arbitrarily. As a corollary, Boyar et al. describe an efficient instantiation for full adders (cf. Sect. 3.3.1.1). In Boyar and Peralta [37]

---

[1] Multiplicative complexity of a function measures the number of AND gates in its circuit (and gives NOT and XOR gates for free).

they show that the multiplicative complexity for computing the Hamming weight of an n-bit string is exactly $n - H^{\mathbb{N}}(n)$ for $n \geq 1$, where $H^{\mathbb{N}}(n)$ is the Hamming weight of n. Recently, Boyar and Peralta [38] proposed a heuristic for optimizing the multiplicative complexity of circuits in a two-step process. The first step reduces the non-linearity, i.e., the number of AND gates, of the circuit; the second step reduces the number of XOR gates in the linear component. This optimization technique is used to construct a compact S-box for the Advanced Encryption Standard (AES) [170] with 32 AND gates and 83 X(N)OR gates [38].

3.1.3 Preliminaries and Notation

Recall, a d-input gate is a boolean function which maps from d input bits to one output bit (cf. Definition 2).

Notation 1 (d-input gate) *We write gates as* $(\text{in}_1, \ldots, \text{in}_d)[T_d]$ *consisting of the list of inputs* $\text{in}_1, \ldots, \text{in}_d$ *and the gate table* T_d, *a string of* 2^d *bits. The i-th element of T_d is equal to the value of the function evaluated by the gate* $G_d(\text{in}_1, \ldots, \text{in}_d)$ *for* $i = \sum_{j=1}^{d} 2^{j-1} \text{in}_j$. *We write 0 as shortcut for the constant (0-input) gate* $()[0]$ *and analogously 1 for* $()[1]$.

Example 1 The gate $(a, b)[0010]$ computes the function $a, b \mapsto (\neg a) \wedge b$.

Definition 3 (*Class and number of d-input gates*) We denote the class of d-input gates with $\mathcal{G}_d := \{(\text{in}_1, \ldots, \text{in}_d)[T_d] : T_d \in \{0, 1\}^{2^d}\}$. The number of d-input gates is

$$\#\mathcal{G}_d = 2^{2^d}. \tag{3.1}$$

Example 2 The class of two-input gates is $\mathcal{G}_2 = \{(a, b)[T_2] : T_2 \in \{0, 1\}^4\}$ and there exist $\#\mathcal{G}_2 = 2^{2^2} = 16$ two-input gates.

Definition 4 (*Trivial d-input gates*) A trivial d-input gate G_d can be expressed as a wire or an equivalent gate with $d' < d$ inputs. A non-trivial gate is not trivial. The class of d-input gates can be partitioned into the class of trivial d-input gates $\mathcal{G}_d^{\text{triv}}$ and the class of non-trivial d-input gates $\mathcal{G}_d^{\text{nontriv}}$, i.e., $\mathcal{G}_d = \mathcal{G}_d^{\text{triv}} \uplus \mathcal{G}_d^{\text{nontriv}}$. From this we have

$$\#\mathcal{G}_d = \#\mathcal{G}_d^{\text{triv}} + \#\mathcal{G}_d^{\text{nontriv}}. \tag{3.2}$$

Notation 2 (Costs of boolean gates and circuits) *We denote the costs of (non-trivial) d-input gates with* $|\mathcal{G}_d|$. *Similarly, the costs of a boolean circuit C, denoted as $|C|$, are the sum of the costs of its gates.*

Example 3 In the point-and-permute GC construction [157] the garbled table of a non-trivial d-input gate has size $|\mathcal{G}_d| = 2^d \cdot (t + 1)$ bits (cf. Sect. 2.2.2).

Assumption 1 (*Monotonic costs of non-trivial gates*) We assume that the costs of non-trivial gates are monotonic, i.e., non-trivial d-input gates have at most the same cost as non-trivial $(d + 1)$-input gates: $|\mathcal{G}_d| \leq |\mathcal{G}_{d+1}|$.

This assumption is fulfilled for the computation and communication costs of all GC constructions presented in Sect. 2.2.2.

3.2   Circuit Optimizations

See Also. Parts of the following results are based on [177, Sect. 8] (Sect. 3.2.1) and [180, Sect. 3] (Sect. 3.2.2).

3.2.1   Minimizing Circuits

In the following we show how the costs of a given boolean circuit can be minimized. Under Assumption 1, the costs of a boolean circuit can be minimized by (1) *replacing trivial gates* with wires or an equivalent gate with fewer inputs, or (2) *removing gates*.

We concentrate on gates with a small number of inputs, i.e., $d \in \{0, 1, 2, 3\}$, as these are the gates of which circuit building blocks for fundamental functionalities such as operations on integer values are composed (cf. Sect. 3.3).

3.2.1.1   Replacing Trivial Gates

Trivial gates can be replaced with a wire or a gate with fewer inputs using the replacement rules shown in Fig. 3.1: The first column (T) of the replacement tables contains the gate table and the second column (To) contains the functionally equivalent replacement in postfix notation. For example the gate $(a, b, c)[T]$ with gate table $T = 00001010$ represents the function $(\neg a) \wedge c$ and hence can be replaced by the functionally equivalent 2-input gate $(a, c)[0010]$. Non-trivial gates which cannot be replaced are not listed or marked with $-$.

The following theorem describes how many of the d-input gates are trivial and hence can be replaced by such rules:

Theorem 1 (Number of trivial d-input gates)

$$\#\mathcal{G}_d^{\text{triv}} = \begin{cases} 0 & \text{for } d = 0, \\ d + \sum_{i=0}^{d-1} \binom{d}{i} \#\mathcal{G}_i^{\text{nontriv}} & \text{for } d > 0. \end{cases} \tag{3.3}$$

$d = 0:$ ()[T]	
T	To
0	-
1	-

$d = 1:$ (a)[T]	
T	To
00	0
01	a
11	1

$d = 2:$ (a,b)[T]			
T	To	T	To
0000	0	1111	1
0011	b	1100	(b)[10]
0101	a	1010	(a)[10]

$d = 3:$ (a,b,c)[T]

T	To	T	To	T	To
00000000	0	01000100	(a,b)[0100]	11111111	1
00000011	(b,c)[0001]	01010000	(a,c)[0100]	11111100	(b,c)[1110]
00000101	(a,c)[0001]	01010101	a	11111010	(a,c)[1110]
00001010	(a,c)[0010]	01011010	(a,c)[0110]	11110101	(a,c)[1101]
00001100	(b,c)[0010]	01011111	(a,c)[0111]	11110011	(b,c)[1101]
00001111	c	01100110	(a,b)[0110]	11110000	(c)[10]
00010001	(a,b)[0001]	01110111	(a,b)[0111]	11101110	(a,b)[1110]
00100010	(a,b)[0010]	10001000	(a,b)[1000]	11011101	(a,b)[1101]
00110000	(b,c)[0100]	10011001	(a,b)[1001]	11001111	(b,c)[1011]
00110011	b	10100000	(a,c)[1000]	11001100	(b)[10]
00111100	(b,c)[0110]	10100101	(a,c)[1001]	11000011	(b,c)[1001]
00111111	(b,c)[0111]	10101010	(a)[10]	11000000	(b,c)[1000]
		10101111	(a,c)[1011]		
		10111011	(a,b)[1011]		

Fig. 3.1 Replacement of trivial d-input gates

Proof As both constant gates are non-trivial we have $\#\mathcal{G}_0^{\text{triv}} = 0$. For d-input gates we have as first d trivial gates the wires $\text{in}_1, \ldots, \text{in}_d$. Additionally, for each subset S_i of $0 \leq i < d$ inputs, each non-trivial i-input gate with inputs S_i can be augmented into a trivial d-input gate G_d where all inputs of G_d which are not in S_i are ignored. $\square$

Using Eqs. (3.1), (3.2) and (3.3), the number of trivial and non-trivial d-input gates can be computed efficiently via dynamic programming. As shown in Table 3.1, the fraction of trivial gates decreases rapidly with d, so for increasing d most gates are non-trivial and cannot be replaced.

3.2.1.2 Removing Gates

To further reduce the circuit size, we describe an optimization algorithm that tries to eliminate constant gates (i.e., 0-input gates) and 1-input gates in a circuit C. The algorithm reduces the number of gates respectively their degree. Besides the well known propagation of constant gates (Step 1), our algorithm additionally eliminates resulting gates with one input by incorporating them into surrounding gates (Step 2 and Step 3), which results in a smaller circuit size.

Terminology. As described in Sect. 2.1.3.1, we assume that the gates within a circuit are ordered topologically. An *output gate* is a gate whose output is also an output of C. Similarly, an *input gate* is a gate which has at least one input that is also an input of C. For gate G_i, $\text{pred}(G_i)$ denotes the set of its predecessors, i.e.,

Table 3.1 Number of trivial d-input gates

d	$\#\mathcal{G}_d^{\text{nontriv}}$	$\#\mathcal{G}_d^{\text{triv}}$	$\#\mathcal{G}_d$	$\#\mathcal{G}_d^{\text{triv}} / \#\mathcal{G}_d$ (%)
0	2	0	2	(0)
1	1	3	4	75
2	10	6	16	37.5
3	218	38	256	14.8
4	64,594	942	65,536	1.44

gates whose output is an input into G_i. Analogously, $\text{succ}(G_i)$ denotes the set of G_i's successors, i.e., gates having the output of G_i as input. The fan-out of a gate G_i is the number of its successors, i.e., $\text{fanout}(G_i) = \#\text{succ}(G_i)$.

Optimization. Algorithm 1 optimizes a circuit C by removing constant and 1-input gates as described in the following:

Algorithm 1 Optimize Circuit by Removing Constant and 1-input Gates

---

Input: circuit C consisting of gates $G_1, .., G_n$ in topological order
  1: **procedure** OPTIMIZE(C)
  2: **Step 1**: Eliminate constant gates
  3:     **for all** constant non-output gates $G_j = ()[v_j]$ **do**
  4:         **for all** gates G_i having the output of G_j as k_i-th input **do**
  5:             eliminateConstInput(G_i, k_i, v_j)
  6:         **end for**
  7:     **end for**
  8: **end Step**
  9: **Step 2**: Eliminate 1-input non-output gates
 10:     **for all** non-output gates G_i with $d_i = 1$ **do**
 11:         integrateInSucc(G_i)
 12:     **end for**
 13: **end Step**
 14: **Step 3**: Eliminate 1-input output gates
 15:     **for all** output gates G_i with $d_i = 1$ **do**
 16:         $\{G_p\} \leftarrow pred(G_i)$
 17:         **if** G_i is not input gate and $\text{fanout}(G_p) = 1$ **then**
 18:             integrateInPred(G_i, G_p)
 19:         **end if**
 20:     **end for**
 21: **end Step**
 22: **end procedure**

- **Step 1: Eliminate constant gates**. The first step of Algorithm 1 eliminates all constant gates $G_j = ()[v_j]$ of circuit C with respective constant value $v_j \in \{0, 1\}$. For all gates G_i with degree d_i having the output of G_j as k_i-th input, the function eliminateConstInput(G_i, k_i, v_j) is invoked that eliminates the corresponding input of G_i. Only the lines of the function table of G_i with value v_j in the k_i-th position are used while the other entries are eliminated, i.e., the modified gate G_i' computes

$G'_i(\text{in}_1, .., \text{in}_{k_i-1}, \text{in}_{k_i+1}, .., \text{in}_{d_i}) = G_i(\text{in}_1, .., \text{in}_{k_i-1}, v_j, \text{in}_{k_i+1}, .., \text{in}_{d_i})$. For an efficient implementation of Algorithm 1 it is crucial that eliminateConstInput() does not copy the entire function table of a gate G_i with degree d_i for each elimination of a constant input as this would result in runtime $\mathcal{O}(\#c_i \cdot 2^{d_i})$ for each gate. Instead, the constant gates are marked in runtime $\mathcal{O}(\#c_i)$ and afterwards all constant gates are eliminated simultaneously in runtime $\mathcal{O}(2^{d_i})$ by copying the corresponding elements of the function table. This results in runtime $\mathcal{O}(2^{d_i})$ per gate. Resulting constant gates G'_i with $d'_i = 0$ are propagated into their successors by recursively calling eliminateConstInput($G_s, k_s, G_i(v_j)$) for all $G_s \in \text{succ}(G'_i)$ having G'_i as k_s-th input. If the constant gate G'_i is not an output gate it is eliminated afterwards. After termination of Step 1 there might be 1-input gates G_i left. The next two steps of Algorithm 1 try to remove these by incorporating their functionalities into their successors (Step 2) or predecessors (Step 3).

- **Step 2: Eliminate non-output gates with one input**. The second step of Algorithm 1 eliminates 1-input gates G_i that are not output gates. The functionality of G_i is incorporated into its successors $G_s \in \text{succ}(G_i)$ by invoking the function integrateInSucc(G_i). This function eliminates G_i by replacing it with a wire and incorporating the functionality of G_i into the function tables of all its successors $G_s \in \text{succ}(G_i)$: Let the output of G_i be the k-th input of G_s and d the degree of G_s. Then, the modified gate G'_s computes $G'_s(\text{in}_1, .., \text{in}_k, .., \text{in}_d) = G_s(\text{in}_1, .., G_i(\text{in}_k), .., \text{in}_d)$. Note that, independent of the functionality of G_i, the resulting gate G'_s has the same number of inputs d as G_s but additionally incorporates the functionality of G_i. As in Step 1, for runtime $\mathcal{O}(2^{d_i})$ per gate the modifications of the function tables are not applied directly but first marked and then done simultaneously.

- **Step 3: Eliminate output gates with one input**. The third step of Algorithm 1 tries to eliminate 1-input gates G_i that are output gates. The functionality of G_i is incorporated into its predecessor G_p, if and only if G_i is the only successor of G_p, i.e., fanout(G_p) $= 1$. In this case, the function integrateInPred(G_i, G_p) is invoked which eliminates gate G_i by replacing it with a wire and incorporates its functionality into gate G_p with d inputs. The modified gate G'_p computes $G'_p(\text{in}_1, .., \text{in}_d) = G_i(G_p(\text{in}_1, .., \text{in}_d))$. As in Step 2, this optimization step is independent of the functionality of G_i and the resulting gate G'_p has the same degree as G_p but additionally incorporates the functionality of G_i.

The following theorem summarizes the correctness and efficiency of Algorithm 1.

Theorem 2 *Algorithm 1 efficiently eliminates all $d \in \{0, 1\}$-input gates that are not output gates of circuit C with size $|C| = n$ in runtime $\mathcal{O}(|C|)$. The optimized circuit C' has at most the same size and computes the same functionality as C.*

Proof Let n denote the number of gates of C in the following.

1. *Termination.* Algorithm 1 always terminates as all loops are upper bounded and the recursive call of eliminateConstInput() in Step 1 terminates if G_i has no successors.
2. *Efficiency.* Step 1 of Algorithm 1 first marks all constant gates in runtime $\mathcal{O}(|C|)$. Afterwards, the marked constant gates are eliminated in $\mathcal{O}(|C|)$. Step 2 also needs at most $\mathcal{O}(|C|)$ operations as first all the successor gates of non-output 1-input gates are marked to incorporate their parent gate in $\mathcal{O}(|C|)$ steps and afterwards the marked gates are eliminated simultaneously in $\mathcal{O}(|C|)$ steps. Analogously, Step 3 runs in $\mathcal{O}(|C|)$ as well. Hence, the overall runtime of Algorithm 1 is in $\mathcal{O}(|C|)$.
3. *All constant gates that are not outputs are eliminated.* Step 1 of Algorithm 1 eliminates all constant gates that are not output gates by incorporating them into their successor gates G_i in eliminateConstInput().
4. *All 1-input gates that are not outputs are eliminated.* Step 2 of Algorithm 1 eliminates all 2-input gates that are not output gates by incorporating them into their successor gates G_i in integrateInSucc().
5. *Size is not increased.* All steps do not increase the size of the circuit: $|C'| \leq |C| = n$.
6. *Functional equivalence.* None of the optimizations performed in Algorithm 1 changes the functionality of C as they incorporate the values of constant gates that are not output gates (Step 1) respectively 1-input gates (Step 2 and Step 3) into the functionality of surrounding gates. The functionality f' computed by the optimized circuit C' with u inputs is identical to the functionality f computed by the original circuit C: $\forall (\text{in}_1, .., \text{in}_u) \in \{0, 1\}^u : f'(\text{in}_1, .., \text{in}_u) = f(\text{in}_1, .., \text{in}_u)$.

This concludes the proof of Theorem 2. $\square$

3.2.2 Minimizing Circuits with Free XOR

As motivated in Sect. 3.1.1, some cryptographic constructions allow evaluation of XOR gates "for free". In the following we show how to optimize circuits when XOR gates have negligible costs compared to non-XOR gates. In this case it is beneficial to minimize the number of non-trivial non-XOR gates while potentially increasing the number of ("free") XOR gates.

Definition 5 (*XOR-trivial d-input gates*) An XOR-trivial d-input gate G_d is either a trivial d-input gate (cf. Definition 4), or it can be replaced by arbitrary many XOR gates and at most one gate with $d' < d$ inputs. XOR-non-trivial gates are not XOR-trivial.

When XOR gates are "for free", an XOR-trivial d input gate has less cost than an XOR-non-trivial d-input gate: $|\mathcal{G}_d^{\text{XORtriv}}| < |\mathcal{G}_d^{\text{XORnontriv}}|$.

XOR-trivial 2-input gates. For $d = 2$ we have as XOR-trivial gates the 6 trivial 2-input gates shown in Fig. 3.1, and additionally the XOR gate with gate table [0110] and the XNOR gate with gate table [1001] which can be replaced by an XOR gate and a NOT gate as $(a, b)[1001] = ((a, b)[0110])[10]$. Overall, the number of XOR-trivial 2-input gates is $\#\mathcal{G}_2^{\text{XORtriv}} = 6 + 2 = 8$ and the fraction of XOR-trivial gates is $\#\mathcal{G}_2^{\text{XORtriv}}/\#\mathcal{G}_2 = 8/16 = 50\,\%$.

XOR-trivial 3-input gates. In the following we show that also half of the 3-input gates are XOR-trivial and can be replaced by at most one 2-input non-XOR gate and at most three XOR gates.

It is easy to see that the optimal topology for replacing a non-trivial 3-input gate (written in postfix notation) is

$$(((\lambda)XOR, (\mu)XOR)[\tau]), \nu)XOR,$$

where $\lambda, \mu, \nu \subseteq \{a, b, c\}$, and τ is the gate table of a 2-input gate. Based on this structure we computed by brute-force enumeration of all $2^3 \cdot 2^3 \cdot 2^{2^2} \cdot 2^3 = 8,192$ possibilities for λ, μ, τ, ν the optimal replacement for XOR-non-trivial 3-input gates.

The resulting 90 replacements for XOR-trivial 3-input gates which are not already listed in the replacement table of 38 trivial gates in Fig. 3.1 are given in Fig. 3.2. Overall, the number of XOR-trivial 3-input gates is $\#\mathcal{G}_3^{\text{XORtriv}} = 90 + 38 = 128$ and the fraction of XOR-trivial 3-input gates is $\#\mathcal{G}_3^{\text{XORtriv}}/\#\mathcal{G}_3 = 128/256 = 50\,\%$.

Before summarizing our main theorem for XOR-trivial gates we introduce the following notation:

Notation 3 (Even and Odd Gates) *An* even *gate is a gate whose truth table T has even parity, i.e., the Hamming weight $H^N(T)$ is even. Otherwise the gate is called* odd.

Example 4 The XOR gate with gate table [0110] and the 3-input gate with gate table [00010111] are even, whereas the OR gate with gate table [0111] is odd.

With this notation we obtain the following theorem for XOR-triviality:

Theorem 3 (XOR-trivial gates) *For $d \in \{2, 3\}$, the XOR-trivial d-input gates are exactly the even gates. Each of them can be replaced by at most one $(d - 1)$-input non-XOR gate and at most 3 XOR gates.*

Proof The proof of Theorem 3 follows from the observations on XOR-trivial 2- and 3-input gates given above and the fact that all 3-input gates listed in Figs. 3.1 and 3.2 are even. □

Further observations on XOR-trivial gates. We observe that for the considered cases $d \in \{2, 3\}$ the fraction of XOR-trivial d-input gates is exactly $50\,\%$, i.e., half of all such d-input gates can be replaced by cheaper ones.

As all practical circuit constructions described in Sect. 3.3 are composed of $d \le 3$-input gates and most of the 3-input gates in these constructions are XOR-trivial, we do not investigate XOR-triviality for $d > 3$ but leave this as an open problem.

```
00000110  ((a,b)[0110],c)[0001]
00001001  ((a,b)[0110],c)[0010]
00010010  (b,(a,c)[0110])[0001]
00010100  (a,(b,c)[0110])[0001]
00010111  (((a,b)[0110],(a,c)[0110])[0001],a)[0110]
00011000  ((a,b)[0110],(a,c)[0110])[0010]
00011011  ((a,(b,c)[0110])[0001],c)[0110]
00011101  ((b,(a,c)[0110])[0001],c)[0110]
00011110  ((a,b)[0001],c)[0110]
00100001  (b,(a,c)[0110])[0100]
00100100  ((a,b)[0110],(a,c)[0110])[0100]
00100111  ((a,(b,c)[0110])[0001],b)[0110]
00101000  (a,(b,c)[0110])[0010]
00101011  (((a,b)[0110],(a,c)[0110])[0010],b)[0110]
00101101  ((a,b)[0010],c)[0110]
00101110  ((b,(a,c)[0110])[0100],c)[0110]
00110101  (((a,b)[0110],c)[0001],b)[0110]
00110110  ((a,c)[0001],b)[0110]
00111001  ((a,c)[0010],b)[0110]
00111010  (((a,b)[0110],c)[0010],b)[0110]
01000001  (a,(b,c)[0110])[0100]
01000010  ((a,b)[0110],(a,c)[0110])[0001]
01000111  ((b,(a,c)[0110])[0001],a)[0110]
01001000  (b,(a,c)[0110])[0010]
01001011  ((a,b)[0100],c)[0110]
01001101  (((a,b)[0110],(a,c)[0110])[0001],c)[0110]
01001110  ((a,(b,c)[0110])[0100],c)[0110]
01010011  (((a,b)[0110],c)[0001],a)[0110]
01010110  ((b,c)[0001],a)[0110]
01011001  ((b,c)[0010],a)[0110]
01011100  (((a,b)[0110],c)[0010],a)[0110]
01100000  ((a,b)[0110],c)[0100]
01100011  ((a,c)[0100],b)[0110]
01100101  ((b,c)[0100],a)[0110]
01101001  ((b,c)[0110],a)[0110]
01101010  ((b,c)[0111],a)[0110]
01101101  ((a,c)[0111],b)[0110]
01101111  ((a,b)[0110],c)[0111]
01110001  (((a,b)[0110],(a,c)[0110])[0001],b)[0110]
01110010  ((a,(b,c)[0110])[0100],b)[0110]
01110100  ((b,(a,c)[0110])[0100],a)[0110]
01111000  ((a,b)[0111],c)[0110]
01111011  (b,(a,c)[0110])[0111]
01111101  (a,(b,c)[0110])[0111]
01111110  ((a,b)[0110],(a,c)[0110])[0111]

11111001  ((a,b)[0110],c)[1110]
11110110  ((a,b)[0110],c)[1101]
11101101  (b,(a,c)[0110])[1110]
11101011  (a,(b,c)[0110])[1110]
11101000  (((a,b)[0110],(a,c)[0110])[1011],b)[0110]
11100111  ((a,b)[0110],(a,c)[0110])[1101]
11100100  ((a,(b,c)[0110])[1101],b)[0110]
11100010  ((b,(a,c)[0110])[1101],a)[0110]
11100001  ((a,b)[1110],c)[0110]
11011110  (b,(a,c)[0110])[1011]
11011011  ((a,b)[0110],(a,c)[0110])[1011]
11011000  ((a,(b,c)[0110])[1101],c)[0110]
11010111  (a,(b,c)[0110])[1101]
11010100  (((a,b)[0110],(a,c)[0110])[1000],a)[0110]
11010010  ((a,b)[1101],c)[0110]
11010001  ((b,(a,c)[0110])[1000],a)[0110]
11001011  (((a,b)[0110],c)[1011],a)[0110]
11001001  ((a,c)[1110],b)[0110]
11000110  ((a,c)[1101],b)[0110]
11000100  (((a,b)[0110],c)[1000],a)[0110]
10111110  (a,(b,c)[0110])[1011]
10111101  ((a,b)[0110],(a,c)[0110])[1110]
10111011  ((b,(a,c)[0110])[1101],c)[0110]
10110111  (b,(a,c)[0110])[1101]
10110100  ((a,b)[1011],c)[0110]
10110010  (((a,b)[0110],(a,c)[0110])[1000],b)[0110]
10110001  ((a,(b,c)[0110])[1000],b)[0110]
10101100  (((a,b)[0110],c)[1011],b)[0110]
10101001  ((b,c)[1110],a)[0110]
10100110  ((b,c)[1101],a)[0110]
10100011  (((a,b)[0110],c)[1000],b)[0110]
10011111  ((a,b)[0110],c)[1011]
10011101  ((a,c)[1011],b)[0110]
10011010  ((b,c)[1011],a)[0110]
10010110  ((b,c)[0110],a)[1001]
10010101  ((b,c)[1000],a)[0110]
10010011  ((a,c)[1000],b)[0110]
10010000  ((a,b)[0110],c)[1000]
10001001  (((a,b)[0110],(a,c)[0110])[1000],c)[0110]
10001101  ((a,(b,c)[0110])[1000],c)[0110]
10001011  ((b,(a,c)[0110])[1000],c)[0110]
10000111  ((a,b)[1000],c)[0110]
10000100  (b,(a,c)[0110])[1000]
10000010  (a,(b,c)[0110])[1000]
10000001  ((a,b)[0110],(a,c)[0110])[1000]
```

Fig. 3.2 Replacement of XOR-trivial 3-input gates (not already listed in Fig. 3.1)

3.3 Efficient Circuit Constructions

In this section we present several frequently used circuit building blocks which benefit from the circuit optimizations of Sect. 3.2 as all occurring 3-input gates are XOR-trivial (= even) and hence can be replaced by at most one 2-input non-XOR gate. The resulting sizes of the circuit constructions described in the following are summarized in Table 3.2.

See Also. Parts of the following results are based [144, Sect. 3] and [109, Sect. 5.1.1] (Sect. 3.3.2).

Table 3.2 Size: Efficient circuit constructions (for n unsigned ℓ-bit values)

Circuit	Standard		Free XOR
#*gates*	2-input	3-input	2-input non-XOR
ADD, SUB (Sect. 3.3.1)	2	$2\ell - 2$	ℓ
ADDSUB (Sect. 3.3.1.3)	$\ell + 1$	2ℓ	ℓ
MUL (Textbook) (Sect. 3.3.2.1)	$\ell^2 + 2\ell - 2$	$2\ell^2 - 4\ell + 2$	$2\ell^2 - \ell$
MUL (Karatsuba) (Sect. 3.3.2.2)			$\approx 9\ell^{1.6} - 13\ell - 34\ell$
CMP (Sect. 3.3.3.1)	1	$\ell - 1$	ℓ
MUX (Sect. 3.3.3.2)	0	ℓ	ℓ
MIN, MAX (Sect. 3.3.3.3)	$n - 1$	$2\ell(n-1) + 2$	$2\ell(n-1) + n + 1$

3.3.1 Addition and Subtraction

3.3.1.1 Addition

An addition circuit (ADD) for adding two unsigned ℓ-bit integers x, y can be efficiently composed from a chain of 1-bit adders (+), often called full-adders, as shown in Fig. 3.3. The first 1-bit adder has constant input $c_1 = 0$ and can be replaced by a smaller half-adder with two inputs. Each 1-bit adder has as inputs the carry-in bit c_i from the previous 1-bit adder and the two input bits x_i, y_i. The outputs are the carry-out bit $c_{i+1} = (x_i \wedge y_i) \vee (x_i \wedge c_i) \vee (y_i \wedge c_i) = (x_i, y_i, c_i)[00010111]$ and the sum bit $s_i = x_i \oplus y_i \oplus c_i = (x_i, y_i, c_i)[01101001]$. All occurring gates are even and can be optimized to a small number of XOR gates. An equivalent construction for computing c_{i+1} with the same number of non-XOR gates was given in [42, 43].

3.3.1.2 Subtraction

Subtraction in two's complement representation is defined as $x - y = x + \neg y + 1$. Hence, a subtraction circuit (SUB) can be constructed analogously to the addition circuit from 1-bit subtractors (−) as shown in Fig. 3.4. Each 1-bit subtractor computes the carry-out bit $c_{i+1} = (x_i \wedge \neg y_i) \vee (x_i \wedge c_i) \vee (\neg y_i \wedge c_i) = (x_i, y_i, c_i))[01001101]$ and the difference bit $d_i = x_i \oplus \neg y_i \oplus c_i = (x_i, y_i, c_i)[10010110]$. The size of SUB is equal to that of ADD.

3.3.1.3 Controlled Addition/Subtraction

A controlled addition/subtraction circuit (ADDSUB) which can add or subtract two unsigned ℓ-bit values x and y depending on a control input bit ctrl can be naturally constructed as a combination of ADD, SUB, and controlled inversion (CNOT) which is equal to (XOR): If ctrl = 1, y is subtracted from x by negating y and adding this to $x + 1$. Otherwise, $x + y + 0$ is output. The resulting circuit is shown in Fig. 3.5.

Fig. 3.3 Circuit: addition
(ADD)

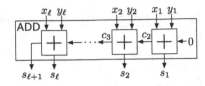

Fig. 3.4 Circuit: subtraction
(SUB)

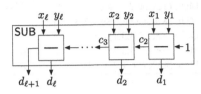

Fig. 3.5 Circuit: controlled
addition or subtraction
(ADDSUB)

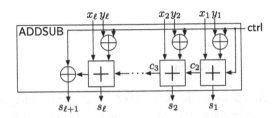

Conversion. The ADDSUB circuit allows conversion back and forth between sign/magnitude and 2's complement representation as follows:

- *sign/magnitude to 2's complement*: A signed integer value x given in sign/magnitude representation with ℓ-bit magnitude can be converted into 2's complement by adding or subtracting the magnitude from 0 depending on the sign of x.
- *2's complement to sign/magnitude*: An $(\ell + 1)$-bit signed integer value x given in 2's complement can be converted into sign/magnitude representation by adding or subtracting the least significant ℓ bits of x from 0 depending on the sign of x, i.e., the most significant bit of x.

3.3.2 Multiplication

We give two circuit constructions for multiplying two ℓ-bit unsigned integer values based on textbook multiplication with size $\mathcal{O}(\ell^2)$ and based on Karatsuba–Ofman multiplication [134] of size $\approx \mathcal{O}(\ell^{1.6})$.

3.3.2.1 Textbook Multiplication

The "textbook method" for multiplying two ℓ-bit unsigned integers x and y multiplies x with each bit of y and adds up all the properly shifted results according to the formula $x \cdot y = \sum_{i=1}^{\ell} 2^{i-1} y_i \cdot x$. This results in a multiplication circuit (MUL) with ℓ^2 1-bit multipliers (2-input AND gates) and $(\ell - 1)$ ℓ-bit adders ADD.

3.3.2.2 Fast Multiplication

As observed by Karatsuba and Ofman [134], multiplication of two large
ℓ-bit unsigned integers can be performed more efficiently using the following
recursive method (details in Algorithm 2): x and y are split into two halves as
$x = x_h 2^{\lceil \ell/2 \rceil} + x_l$ and $y = y_h 2^{\lceil \ell/2 \rceil} + y_l$. Then, the product can be computed as
$x \cdot y = (x_h 2^{\lceil \ell/2 \rceil} + x_l)(y_h 2^{\lceil \ell/2 \rceil} + y_l) = z_h 2^{2\lceil \ell/2 \rceil} + z_d 2^{\lceil \ell/2 \rceil} + z_l$. After computing
$z_h = x_h y_h$ and $z_l = x_l y_l$, z_d can be computed with only one multiplication as
$z_d = (x_h + x_l)(y_h + y_l) - z_h - z_l$. This process is continued recursively until
the numbers are smaller than some threshold θ ($\theta = 19$ in the case of free XOR as
described below) and multiplied with the classical textbook method.

Algorithm 2 Karatsuba Multiplication

Input: x, y (ℓ-bit unsigned integers), constant threshold θ
Output: $x \cdot y$ (2ℓ-bit unsigned integer)
 1: **function** KARATSUBA(x, y)
 2:　　**if** $\ell \leq \theta$ **then**
 3:　　　　**return** TEXTBOOK(x, y)
 4:　　**end if**
 5:　　$x_h \| x_l \leftarrow x$ 　　　　　　　　　　　　　　　　$\triangleright x = x_h 2^{\lceil \ell/2 \rceil} + x_l$
 6:　　$y_h \| y_l \leftarrow y$ 　　　　　　　　　　　　　　　　$\triangleright y = y_h 2^{\lceil \ell/2 \rceil} + y_l$
 7:　　$p_h \leftarrow$ KARATSUBA(x_h, y_h)
 8:　　$p_l \leftarrow$ KARATSUBA(y_l, y_l)
 9:　　$x_s \leftarrow x_h + x_l$
10:　　$y_s \leftarrow y_h + y_l$
11:　　$p_s \leftarrow$ KARATSUBA(x_s, y_s)
12:　　$p_d \leftarrow p_s - p_h - p_l$
13:　　**return** $p_h 2^{2\lceil \ell/2 \rceil} + p_d 2^{\lceil \ell/2 \rceil} + p_l$
14: **end function**

Overall, multiplying two ℓ-bit numbers with Karatsuba's method requires three
multiplications of $\ell/2$-bit numbers and some additions and subtractions with linear
bit complexity resulting in costs

$$T_{Kara}(\ell) = \begin{cases} \mathcal{O}(\ell^2) & \text{if } \ell \leq \theta, \\ 3T_{Kara}(\ell/2) + c\ell + d & \text{else} \end{cases}$$

for constants c and d. The master theorem [64, Sect. 4.3f] yields asymptotic complexity $T_{Kara}(\ell) \in \mathcal{O}(\ell^{\log_2 3}) \approx \mathcal{O}(\ell^{1.585})$.

3.3.2.3 Multiplication Circuit Complexity

In TASTY (cf. Sect. 5.2) we have implemented both methods for multiplication
using the efficient addition and subtraction circuits of Sect. 3.3.1 optimized for a

small number of non-XOR gates. When XOR gates are "for free", we experimentally determined the optimal threshold to be $\theta = 19$, i.e., Karatsuba multiplication is more efficient than textbook multiplication already for multiplication of 20 bit operands (cf. Fig. 3.6 and Table 3.3).

By interpolating through the points for bit length $\ell \in \{32, 64, 128\}$ and solving the resulting system of linear equations we obtain as an approximation for the number of non-XOR gates

$$T_{Kara}(\ell) \approx \begin{cases} 2\ell^2 - \ell & \text{for } \ell \leq 19 \\ 9.0165\ell^{1.585} - 13.375\ell - 34 & \text{else.} \end{cases}$$

3.3.3 Comparison, Minima and Maxima

3.3.3.1 Comparison

Comparing two ℓ-bit unsigned integers x and y is equivalent to computing the function

$$z = [x > y] := \begin{cases} 1 & \text{if } x > y, \\ 0 & \text{else.} \end{cases}$$

As shown in Fig. 3.8, a comparison circuit (CMP) can be composed from ℓ sequential 1-bit comparators ($>$). The first 1-bit comparator has constant input $c_1 = 0$ and can be replaced by a 2-input gate. Each 1-bit comparator has as inputs the carry-in bit c_i from the previous 1-bit comparator and the two input bits x_i, y_i. Its output is the carry-out bit $c_{i+1} = (x_i \wedge y_i) \vee (x_i \wedge c_i) \vee (y_i \wedge c_i) = (x_i, y_i, c_i)[01001101]$.

Note that this improved bit comparator computes the same function as the carry output of the subtraction circuit of Sect. 3.3.1.2: as $[x > y] \Leftrightarrow [x - y - 1 \geq 0]$, this coincides with an underflow in the corresponding subtraction denoted by the subtractor's most significant output bit $d_{\ell+1}$.

Modifications. Comparison circuits for $[x < y]$, $[x \geq y]$, or $[x \leq y]$ can be obtained from the comparison circuit for $[x > y]$ by interchanging x with y and/or setting the initial carry to $c_1 = 1$.

A circuit to check if two values are equal ($[x = y]$), can be constructed analogously by computing $c_{i+1} = (x_i = y_i) \wedge c_i = (x_i, y_i, c_i)[00001001]$ with $c_1 = 1$. Similarly, a circuit for checking inequality ($[x \neq y]$) computes $c_{i+1} = (x_i \neq y_i) \vee c_i = (x_i, y_i, c_i)[01101111]$ with $c_1 = 0$. These circuits for (in-)equality have the same size as the circuit constructions given in Kolesnikov and Schneider [142].

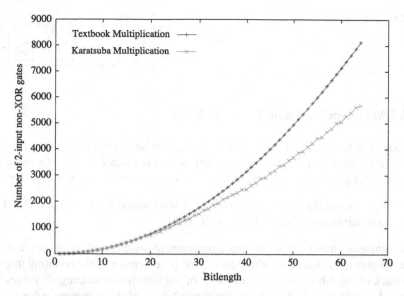

Fig. 3.6 Size: multiplication circuits (with free XOR)

Table 3.3 Size: multiplication circuits (in # 2-input non-XOR gates)

Bitlength ℓ	19	20	32	64	128
Textbook	703	780	2,016	8,128	32,640
Karatsuba	703	721	1,729	5,683	17,973
Improvement (%)	0.0	7.6	14.2	30.1	44.9

Fig. 3.7 Circuit: multiplexer (MUX)

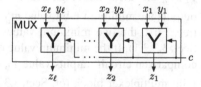

3.3.3.2 Multiplexer

An ℓ-bit multiplexer circuit MUX selects its output z to be its left ℓ-bit input x if the input selection bit c is 0, respectively its right ℓ-bit input y otherwise. As shown in Fig. 3.7, this circuit can be composed from ℓ parallel 1-bit multiplexers (Y), where Y computes $z_i = (x_i, y_i, c)[01010011]$. After optimization, this circuit has the same size as the construction given in [142].

Fig. 3.8 Circuit: comparison (CMP)

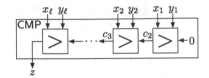

3.3.3.3 Minimum/Maximum Value and Index

We show how to combine the blocks for comparison and multiplexing into a circuit MIN which selects the minimum value m and minimum index i of a list of n unsigned ℓ-bit integers $x_0, \ldots, x_{n-1}$, i.e., $\forall j \in \{0, \ldots, n-1\} : (m < x_j) \vee (m = x_j \wedge i \leq j)$.

Example 5 For the list $3, 2, 5, 2$ the output of MIN would be $m = 2$ and $i = 1$ as the leftmost minimum value of 2 is at position 1.

To minimize circuit size we arrange a tournament-style circuit so that some of the index wires can be reused and eliminated. That is, at depth d of the resulting tree we keep track of the ℓ-bit minimum value $m^{\ell 2}$ of the sub-tree containing 2^d values but store and propagate only the d least significant bits i_d^d of the minimum index.

W.l.o.g. we assume that n is a power of two in the following and hence the minimum index can be represented with exactly $\log n$ bits. (If n is not a power of two, the circuit given in the following can be constructed for the next power of two and optimized afterwards.) More specifically, the minimum value and minimum index are selected pairwise in a tournament-like way using a tree of minimum blocks (min) as shown in Fig. 3.9. As shown in Fig. 3.9b, each minimum block at depth d gets as inputs the minimum ℓ-bit values $m_{d,L}^\ell$ and $m_{d,R}^\ell$ of its left and right subtrees T_L, T_R and the d least significant bits of their minimum indices $i_{d,L}^d$ and $i_{d,R}^d$, and outputs the minimum ℓ-bit value m_{d+1}^ℓ and $(d+1)$-bit minimum index i_{d+1}^{d+1} of the tree. First, the two minimum values are compared with a comparison circuit (cf. Sect. 3.3.3.1). If the minimum value of T_L is bigger than that of T_R (in this case, the comparison circuit outputs value 1), m_{d+1}^ℓ is chosen to be the value of T_R with an ℓ-bit multiplexer block (cf. Sect. 3.3.3.2). In this case, the minimum index i_{d+1}^{d+1} is set to 1 concatenated with the minimum index of T_R using another d-bit multiplexer. Alternatively, if the comparison yields 0, the minimum value of T_L and the value 0 concatenated with the minimum index of T_L are output.

Overall, the size of the efficient minimum circuit is $|\mathsf{MIN}^{\ell,n}| = (n-1) \cdot (|\mathsf{CMP}^\ell| + |\mathsf{MUX}^\ell|) + \sum_{j=1}^{\log n} \frac{n}{2^j} |\mathsf{MUX}^{j-1}|$. When XOR gates are free, the resulting number of non-XOR gates is $(n-1) \cdot (\ell + \ell) + n \sum_{j=1}^{\log n} \frac{j-1}{2^j} < 2\ell(n-1) + n(1 + \frac{1}{n}) = 2\ell(n-1) + n + 1$.

Modifications. A circuit MAX which computes the maximum value and index can be constructed similarly by constructing max blocks which use $\mathsf{CMP}_<$ instead of $\mathsf{CMP}_>$. When only the minimum (maximum) value or only the minimum (maximum) index is needed, the corresponding multiplexers can be omitted.

---

[2] To simplify presentation we write the number of bits of a variable as superscript index.

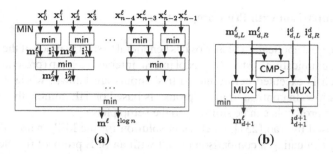

Fig. 3.9 Circuit: minimum (MIN). **a** Circuit Structure: Minimum (MIN); **b** Block: Minimum (min)

3.4 Applications: Secure Comparison and Auctions

The efficient circuit building blocks described in Sect. 3.3 with optimizations for free XORs can serve as a basis for highly efficient protocols for secure comparison (Sect. 3.4.1) and auction protocols (Sect. 3.4.2) as described next.

See Also. Parts of the following results are based on [144, Sect. 5].

3.4.1 Comparison (Millionaires Problem)

The "Millionaires Problem" was introduced by Yao in [230] as motivation for secure computation: two millionaires want to securely compare their respective private input values (e.g., their amount of money) without revealing more information than the outcome of the comparison to the other party. More concretely, client C holds a private ℓ-bit value x and server S holds a private ℓ-bit value y. The output bit $z = [x < y]$ should be revealed to both. In the semi-honest case it is sufficient to consider the case where C obtains the output and forwards it to S.

An example application that can be reduced to multiple invocations of secure comparison is the secure computation of the kth-ranked element of the union of two datasets held by two parties [1].

When instantiated efficiently, Yao's GC protocol (cf. Sect. 2.3.1.1) is the most efficient solution to solve the Millionaires Problem for computationally bounded players. Many other protocols for secure comparison based on HE were proposed which work in a stronger model, where one player is computationally unbounded [29, 78], or extended the Millionaires Problem to other scenarios such as online auctions [66–68] (cf. Sect. 3.4.2.2), or multiparty computation [89].

In the following, we show that evaluating our improved comparison circuit of Sect. 3.3.3.1, using an efficient GC-based SFE protocol with free XORs, is more efficient than previous comparison protocols.

3.4.1.1 Comparison with Pre-Computations

In many practical application scenarios it is beneficial to shift as much of the computation and communication cost of a protocol into an interactive setup (pre-computation) phase, which is executed before the parties' inputs are known, while the parties' workload is low. In contrast to many protocols based on HE, almost the entire cost of GC-based protocols can be shifted into the setup phase.

As proposed by Yao [231], an efficient solution for the Millionaires Problem is obtained by evaluating a comparison circuit with an SFE protocol (cf. Sect. 2.3.1), where C provides her ℓ-bit input x, S provides his ℓ-bit input y, and C obtains the output bit $z = [x < y]$. We instantiate this protocol with the following efficient primitives in the Random Oracle (RO) model: As a comparison circuit we use our optimized comparison circuit of Sect. 3.3.3.1 with ℓ 2-input non-XOR gates. As an efficient GC protocol we use garbled row reduction [164] with free XORs [142] (cf. Sect. 2.2.3). Hence, the resulting GC has size $\ell \cdot 3(t + 1)$ bits. For Oblivious Transfer (OT) we use the protocol of Naor and Pinkas [163] with pre-computations of Beaver [23] (cf. Sect. 2.2.3).

3.4.1.2 Complexity Evaluation

In the following we show that the GC-based comparison protocol has a smaller complexity than previous HE-based protocols.

Computation Complexity. As our improved circuit for integer comparison described in Sect. 3.3.3.1 consists of ℓ non-XOR 2-input gates, C needs to invoke the underlying cryptographic hash-function (e.g., SHA-256 for $t = 128$ bit symmetric security) exactly ℓ times to evaluate the GC (cf. Sect. 2.2.2.4). All other operations are negligible (XOR of bitstrings). Hence, the computational complexity of the online phase of our protocol is negligible compared to that of protocols based on HE. Even with an additional setup phase, those HE-based protocols need to invoke a few modular operations in the online phase for each input bit which are usually more expensive by several orders of magnitude than the evaluation of a cryptographic hash function used in our protocols. For large ℓ or parallel comparisons, our protocols have smaller overall computation complexity than comparison protocols based on HE, as the number of public-key operations is independent of the bitlength ℓ when using the efficient OT extension of Ishai et al. [121] as described in Sect. 2.2.3.

Communication Complexity. In Table 3.4 we compare the communication complexity of our GC-based comparison protocol with that of previous HE-based protocols as described in the following. The names and sizes of the security parameters are chosen according to Sect. 2.1.1.2.

In our protocol, the communication complexity of the setup phase consists of $\ell \cdot 3(t + 1)$ bits for sending the GC, $6\ell t$ bits for pre-computing OT (using the OT protocol of Naor and Pinkas [163] over Elliptic Curves (EC)), and ℓ bits for allowing C to decrypt the output, i.e., asymptotically $9\ell t$ bits. The communication complexity of the online phase consists of $\ell(t + 1)$ bits for sending S's garbled inputs $\tilde{y}$, and

Table 3.4 Communication complexity: comparison protocols (on ℓ-bit values)

Protocol	Asymptotic Complexity	[ℓByte] for security level			
		ultra-short	short	medium	long
[78]	$(\kappa + 1)\ell T$	6, 396	9, 102	12, 464	16, 646
[29] + [176]	$4\ell T$	624	888	1, 216	1, 624
[29] + [66]	$2\ell T$	312	444	608	812
[29] + EC-[75]	$8\ell t$	80	96	112	128
GC-based comparison					
Setup phase	$9\ell t$	90	108	126	144
Online phase	$3\ell t$	30	36	42	48
Combined	$10\ell t$	100	120	140	160

$\sim 2\ell t$ bits for the online OT phase for $\mathcal{C}$'s garbled inputs $\tilde{y}$, i.e., asymptotically $3\ell t$ bits. When pre-computations are not possible, the two phases can be combined and OT need not be pre-computed.

The comparison protocol of Fischlin [78] uses the Goldwasser-Micali XOR HE scheme [100] and has asymptotic communication complexity $\ell T(\kappa + 1)$, where κ is the statistical correctness parameter (e.g., $\kappa = 40$). The comparison protocol of Blake and Kolesnikov [29] requires the transfer of 2ℓ ciphertexts of an additively HE scheme which can be instantiated with any of the schemes presented in Sect. 2.2.1.1: Paillier [176] (as proposed in [29]), Damgård-Geisler-Krøigaard (DGK) [66–68] (as used for example in [76]), or lifted EC-ElGamal [75] (as used for example in [21]).

Comparison with pre-computations. As shown in Table 3.4, the communication complexity of the online phase of our improved GC-based comparison protocol is substantially lower than that of previous HE-based protocols. The GC-based protocol improves over the online communication complexity of the best HE-based protocol ([29] + EC-ElGamal [75]) by a factor of $\frac{8\ell t}{3\ell t} \approx 2.7$ times.

Comparison in parallel and without pre-computations. We note that when many comparisons are performed in parallel, the communication complexity of our protocol is improved by $2\ell t$ as OTs can be extended efficiently (cf. Sect. 2.2.3.4): the setup phase has asymptotic communication complexity $7\ell t$; the combined protocol without pre-computations has $8\ell t$ communication. In this case, even if the application scenario does not allow pre-computations, the communication complexity of our GC-based comparison protocol is similar to that of the best HE-based comparison protocol of Blake and Kolesnikov [29] instantiated with EC-ElGamal [75].

3.4.2 Auctions

In standard auction systems such as eBay, the auctioneer learns the inputs of all bidders and hence can deduce valuable information about the bidding behavior of unsuccessful bidders or cheat while computing the auction function depending on bidders' input values. To overcome this, a secure protocol can be used instead. Bidders provide their bids in "encrypted" form to the protocol which allows the auctioneer to compute the auction function without learning the bids.

Next, we show how our efficient circuit constructions can be used to improve two previously proposed secure auction systems: in *offline auctions* all bids are collected before the auction function is computed (Sect. 3.4.2.1), whereas in *online auctions* the bids are input dynamically and the current highest bid is published (Sect. 3.4.2.2).

3.4.2.1 Offline Auctions

The offline auction system of Naor et al. [164] evaluates the auction function in the SMPC setting with two servers as described in Sect. 2.3.2.1: The *auction issuer* creates the GC and the *auctioneer* evaluates it on the garbled bids provided by the bidders. Finally, the auctioneer publishes the outcome of the auction which can be verified with the help of the auction issuer.

In order to run a first-price auction which outputs the maximum bid and the index of the maximum bidder, a circuit for the maximum value and its index (cf. Sect. 3.3.3.3) needs to be evaluated. The GC for first-price auctions with n bidders and ℓ-bit bids used in [164] requires approximately $15n\ell$ table entries when using the GC method of [164] without free XOR (cf. Sect. 2.2.2.3). Instead, our improvements result in a GC with approximately $3 \cdot 2\ell n = 6\ell n$ table entries, when evaluating our optimized maximum circuit presented in Sect. 3.3.3.3 ($\approx 2\ell n$ non-XOR gates) with the GC method of [164] and free XORs. The communication between the two computation servers for sending the GC as well as the auction issuer's time for creating the GC are improved by a factor of $\frac{15}{6} = 2.5$. The time for evaluating the GC is reduced as well.

We note that also other types of auctions, such as second-price auctions, benefit from our improved circuit constructions as these auction functions are also composed of comparisons and multiplexers for which we give more efficient circuit constructions with free XORs.

3.4.2.2 Online Auctions

In an on-line auction system, each bidder submits a maximum bid to the system and goes offline afterwards. The system automatically bids for each bidder until their respectively specified maximum is exceeded. In contrast to offline auctions, new maximum bids can be submitted dynamically until the auction ends, e.g., at a specified time. Clearly, the maximum bid is confidential information as both the auctioneer and other bidders could exploit this information to their advantage.

The secure online auction system proposed in [66–68] extends the idea of splitting the computation of the auction function between two servers, the auctioneer (called the *server*) and another party (called the *assisting server*) who are assumed not to collude. Each bidder secret-shares her bid and sends it to both servers over a secure channel. The two servers dynamically compare the secret-shared maximum bids with the public value of the current highest bid. The output of the comparison is public and determines whether the bidder is still in the game and wants to raise the bid, say by some fixed amount agreed in advance. The secure comparison guarantees that neither of the two servers learns any information on the maximum bid other than the comparison result. A detailed description of the scenario can be found in [68].

HE-based Online Auctions. The protocol proposed by [66–68] is a modification of an HE-based comparison protocol, where one input is bitwise secret-shared between the two parties. The communication complexity of this protocol is the same as for the HE-based comparison protocols of Sect. 3.4.1.

GC-based Online Auctions. We propose to use the efficient GC-based comparison protocol of Sect. 3.4.1 for the online auctions with inputs given in different forms: the maximum bid is secret-shared between both players (see below for a simple technique to use such inputs in GC) and the other input is publicly known to both parties (e.g., can be treated as a private input of the circuit constructor S). The resulting GC-based protocol for online auctions has exactly the same performance as our solution for the Millionaires Problem described in Sect. 3.4.1.

Performance Comparison. As the complexity of the HE- and GC-based online auction protocols is similar to the complexity of the corresponding comparison protocols of Sect. 3.4.1, the GC-based protocol outperforms the HE-based protocol for the same reasons. In particular, the possibility to move all expensive operations into the setup phase, which can be executed during idle times (whenever no new bids are received), is very beneficial for the online auctions scenario as this enables the bidders to instantly see if their current bid was successful or if another bidder meanwhile gave a higher bid. This feature is important especially towards the end of the auction, where the frequency of bids is usually very high.

Secret-Shared Inputs. As proposed in [77], a bit b can be secret-shared between C holding share b_C and S holding share b_S, with $b = b_C \oplus b_S$. A secret-shared input bit b can be converted into a garbled input $\widetilde{b}$ for C using a 1-out-of-2 OT protocol (cf. Sect. 2.2.2.3): C inputs b_C and S inputs the two corresponding garbled values in the usual order $\widetilde{b}^0, \widetilde{b}^1$ if $b_S = 0$ or swaps them to $\widetilde{b}^1, \widetilde{b}^0$ otherwise. It is easy to verify that C obliviously obtains the correct garbled value $\widetilde{b}$ for the shared bit b.

Chapter 4
Hardware-Assisted Garbled Circuit Protocols

4.1 Creating Garbled Circuits with Hardware Token

The techniques described in this section allow one to make the communication complexity of Garbled Circuit (GC)-based Secure Function Evaluation (SFE) protocols *independent* of the size of the evaluated functionality. Instead of generating and transferring the GC over the network, we propose to use a low-cost tamper-proof Hardware (HW) token which is issued and trusted by one party and generates the GC on behalf of her. The token is cheap as it executes only symmetric-key operations (e.g., SHA and AES) and has small constant-size RAM (much smaller than the size of the circuit), but we do not resort to implementing expensive secure external RAM. We provide two solutions; in one, $\mathcal{T}$ keeps the state in secure non-volatile storage (a monotonic counter), while in the other, $\mathcal{T}$ maintains no long-term state.

See Also. Parts of the following results are based on [127, 128].

4.1.1 Motivation and Setting

The communication complexity of GC-based protocols is dominated by the transfer of the GC $\widetilde{C}$, as for each (non-XOR) gate, a garbled table has to be sent (cf. Sect. 2.2.2.4), e.g., the GC for AES has size 0.5 MB [180]. Further, if security against more powerful adversaries is required; the use of the standard cut-and-choose technique implies transfer of multiple GCs (cf. Sect. 2.3.1.2). While transmission of this large amount of data is possible for exceptional occurrences, in most cases the network will not be able to sustain the resulting traffic. This holds especially for larger-scale deployment of secure computations, e.g., by banks or service providers, with a large number of customers. Additional obstacles include the energy

T. Schneider, *Engineering Secure Two-Party Computation Protocols*,
DOI: 10.1007/978-3-642-30042-4_4, © Springer-Verlag Berlin Heidelberg 2012

consumption required to transmit/receive the data, and the resulting reduced battery life in mobile clients, such as smartphones.[1]

In this section we show how to remove this expensive communication requirement by generating the GC *locally* with a secure HW token T. The token is issued by the GC creator S and given to the GC evaluator C. C communicates locally with T, and remotely with S, to obtain the GC. There is no direct channel between T and S, but C can pass (and potentially interfere with) messages between T and S. T is created by S, so S trusts T; however, as C does not trust S, she also does not trust the token T to behave honestly.[2]

Hardware assumption. We assume T is tamper-proof. We argue that this assumption is reasonable. Indeed, while every token can likely be broken into, given sufficient resources (see for example the physical breach of the Trusted Platform Module (TPM) [114]), we are motivated by the scenarios where the payoff of the break is far below the cost of the break. This holds for relatively low-value transactions such as cell phone or TV service, where the potential benefit of the attack (e.g., free TV for one user) is not worth the investment of thousands or tens of thousands of dollars to break into the card. For higher-value applications the cost of the attack are raised by using a high-end token T, e.g., a smart card certified at FIPS 140-2, level 3 or 4.

Hardware restrictions. As we assume the token to be produced in large quantities, we try to minimize its costs (e.g., chip surface) and make the assumptions on it as weak as possible. In particular our token requires only restricted computational capabilities (no public-key operations) and a small, constant amount of secure RAM. We consider T with and without a small, constant amount of secure non-volatile storage.

4.1.2 Related Work

Related work on using tokens for secure computations can be divided into the following three categories, summarized in Table 4.1.

(A) Setup assumptions for the Universal Composability (UC) framework. As shown in [56], UC-secure SFE protocols can be constructed from UC commitments. In turn, UC commitments can be constructed from signature cards trusted by both parties [111], or from tamper-proof tokens created and trusted only by the issuing party [59, 70, 135, 161]. Here, [59] consider stateless tokens, and [161] require only one party to issue a token. This line of research mainly addresses the feasibility of UC-secure computation based on tamper-proof HW and relies on expensive primitives such as generic zero-knowledge proofs. Our token-based SFE protocols do not achieve UC-security, but are far more practical.

---

[1] In some cases, the impact can be mitigated by creating and transferring GCs in the setup phase. However, this is not fully satisfactory. Firstly, even more data needs to be transferred since demand cannot be perfectly predicted. Further, this creates other problems, such as requiring large long-term storage on client devices.

[2] Note, if C in fact trusts T to behave honestly, then there exists a trivial solution, where C would let T compute the function on her inputs [117].

Table 4.1 Secure protocols using hardware tokens

	Reference	Functionality	N	Trust	Stateful	Public-key operations
(A)	[111]	UC commitment	2	Both	Yes	Yes
	[70, 135]	UC commitment	2	**Issuer**	Yes	Yes
	[59]	UC commitment	2	**Issuer**	**No**	Yes
	[161]	UC commitment	1	**Issuer**	Yes	**No**
(B)	[108]	Set intersection, ODBS	1	Both	Yes	**No**
	[106]	Non-interactive OT	1	Both	Yes	Yes
	[211]	Verif. Enc., Fair Exch.	1	Both	Yes	Yes
	[141]	OT	1	**Issuer**	**No**	**No**
	[79]	Set intersection	1	**Issuer**	Yes	**No**
	[73]	Non-interactive OT	1	**Issuer**	Yes	**No**
(C)	[80]	**SFE**	2	Both	Yes	Yes
	[116, 118, 119]	**SFE**	1	Both	Yes	Yes
	This work	**SFE**	1	Issuer	Yes/No	**No**

Columns denote the number of tokens N, who trusts the token(s), if token(s) are stateful or stateless, and perform public-key operations. Properties more desired for practical applications are in bold font

(B) Efficiency Improvements for Specific Functionalities. Efficient protocols with a tamper-proof token trusted by both players have been proposed for specific functionalities such as set intersection and Oblivious Database Search (ODBS) [108], non-interactive Oblivious Transfer (OT) [106], and verifiable encryption and fair exchange [211]. Protocols secure against covert adversaries with one token trusted by its issuer only were proposed for interactive OT [141] (stateless token) and set intersection [79] (stateful token). Non-interactive OT with security against malicious adversaries with a stateful token was proposed recently in [73]. In contrast, we solve the general SFE problem.

(C) Efficiency Improvements for Arbitrary Functionalities. Clearly, SFE is efficient if aided by one or multiple Trusted Third Parties (TTPs), that simply compute(s) the function: SFE aided by multiple smartcards as TTPs was considered in [80]; the Faerieplay project uses a cryptographic co-processor as a single TTP [116, 118, 119]. The protocol of [110] provides SFE with aborts secure against malicious adversaries where only one party obtains the output. Their construction uses a TTP that generates the GC and guarantees its well-formedness with a signature. By adapting our techniques of Sect. 4.1.4, the TTP can be implemented with constant memory within a HW token. In contrast to these protocols we do not use any TTP; our token is only trusted by its issuer.

Fig. 4.1 Model overview:
token-assisted SFE

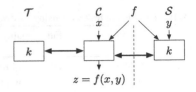

4.1.3 Architecture, System and Trust Model

We present in detail our setting, players, and HW and trust assumptions.

As shown in Fig. 4.1, we extend the two-party SFE setting consisting of client C and server S with a third party, the tamper-proof token T, issued and trusted by S and given to C. Our goal is to let C and S evaluate the public function f on their respective private inputs x and y with T's help.

Communication. $C \leftrightarrow S$: We view this as an expensive channel. Communication $C \leftrightarrow S$ flows over the Internet, and may include a wireless or cellular link. This implies small link bandwidth and power consumption concerns of mobile devices. We wish to minimize the utilization of this channel.

$T \leftrightarrow C$: As T is held locally by C, this is a cheap channel (both in terms of bandwidth and power consumption), suitable for transmission of data linear in the size of f.

$T \leftrightarrow S$: There is no direct channel between T and S, but, of course, C can pass (and potentially interfere with) messages between T and S.

Trust. $C \leftrightarrow S$: As in the standard SFE scenario, C and S don't trust each other. We address semi-honest, covert, and malicious C and S.

$S \leftrightarrow T$: T is fully trusted by S, as T is tamper-proof. S and T share a secret key k to establish a secure channel and to derive joint randomness.

$T \leftrightarrow C$: C does not trust T, as T is the agent of S, and may communicate with S through covert channels.

Storage, computation and execution. C and S are computationally strong devices which can perform both symmetric- and asymmetric-key operations.[3] Both have sufficient memory, linear in the size of f. C has control over T, and can reset it, e.g., by interrupting its power supply. As justified in Sect. 4.1.1, T is a cheap special purpose HW with minimum chip surface: T has circuitry only for evaluating symmetric-key primitives in HW (no public-key or true random number generator) and has a small secure RAM. It may (Sect. 4.1.4.3) or may not (Sect. 4.1.4.4) have small non-volatile secure storage,[4] unaffected by the resets by C.

---

[3] If needed, C's capabilities may be enhanced by using a HW accelerator (cf. Sect. 4.2.3).

[4] T's key k is a fixed part of its circuit, and is kept even without non-volatile storage.

4.1.4 Token-Assisted SFE

We start with a high-level overview of our protocols (Sect. 4.1.4.1) and present the technical details of our construction afterwards; efficient circuit representation (Sect. 4.1.4.2), and GC generation by stateful (Sect. 4.1.4.3) and stateless tokens (Sect. 4.1.4.4). Finally, we give implementation results in Sect. 4.1.4.5.

4.1.4.1 Protocols Overview

Our constructions are a natural (but technically involved) modification of the Software (SW)-based SFE protocols described in Sect. 2.3.1 that split the actions of the server into two parts (now executed by S and T) while maintaining provable security. We offload most of the work (notably, GC generation) to T, thus achieving important communication savings, and partially offloading S's computation to T.

We start our discussion with the solution in the semi-honest model. However, as our modifications of the basic algorithms for GC creation are secure against malicious actions, our protocols are easily and efficiently extendible to covert and malicious settings.

At the high level, our token-assisted SFE protocols work as shown in Fig. 4.2: C obtains the garbled inputs $\tilde{x}, \tilde{y}$ from S, and the GC $\tilde{f}$ corresponding to the function f from T. Then, C evaluates $\tilde{f}$ on $\tilde{x}, \tilde{y}$ and obtains the result $z = f(x, y)$.

It is easy to see that the introduction of T and offloading to it some of the computation does not strengthen S, and thus does not bring security concerns for C (as compared to standard two-party SFE). On the other hand, separating the states of S and T, placing C in control of their communication, and C's ability to reset T introduces attack opportunities for C. We show how to address these issues with the proper synchronization and checks performed by S and T.

Our main tool is the use of a unique session identifier sid for each GC evaluation. From sid and the shared secret key k, S and T securely derive a session key K, which is then used to derive the randomness used in GC generation. We emphasize that each token shares a different random key k with S to obtain unique session keys K and to avoid break-one-break-all behavior of multi-token systems. Jumping ahead (details in Sect. 4.1.4.3), we note that sid uniqueness is easily achieved if T is stateful simply by setting sid equal to the value of a strictly monotonic session counter ctr maintained by T. However, if T is stateless, C can always replay S's messages. In Sect. 4.1.4.4 we show how to ensure that replays do not help C.

Since S and T derive the same randomness for each session, the (same) GC $\tilde{f}$ can be generated by T. Unfortunately, the weak T cannot store the entire function f. Instead, C provides the circuit corresponding to function f gate-by-gate to T, and obtains the corresponding garbled gate of $\tilde{f}$. The garbled gate can immediately be evaluated by C and need not be stored. C is prevented from providing a wrong f to T, as follows. First, S issues a Message Authentication Code (MAC) of f, e.g., mac = $\text{MAC}_k(\text{sid}, f)$, where f is the agreed circuit representation of the evaluated function

Fig. 4.2 Protocols overview: token-assisted SFE

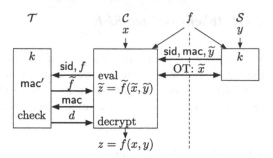

(cf. Sect. 4.1.4.2). Further, $\mathcal{T}$ computes its version of the above MAC, mac', as it answers $\mathcal{C}$'s queries in computing $\widehat{f}$. Finally, $\mathcal{T}$ reveals the decryption information d that allows $\mathcal{C}$ to decrypt the output wires only if $\mathcal{C}$ provides the matching mac.

Garbled Inputs. The garbled input $\widetilde{y}$ of $\mathcal{S}$ can be computed by $\mathcal{S}$ and sent to $\mathcal{C}$, requiring $|y| \cdot (t + 1)$ bits communication, where t is the symmetric security parameter (cf. Sect. 2.1.1.2). Alternatively, if $\mathcal{T}$ is stateful, $\mathcal{S}$ can establish a secure channel with $\mathcal{T}$, e.g., based on session key K, send y over the channel, and have $\mathcal{T}$ output $\widetilde{y}$ to $\mathcal{C}$. This achieves the optimal communication between $\mathcal{S}$ and $\mathcal{C}$ of $\Theta(|y|)$ bits.

The garbled input $\widetilde{x}$ corresponding to $\mathcal{C}$'s input x can be transferred from $\mathcal{S}$ to $\mathcal{C}$ with a parallel OT protocol (cf. Sect. 2.2.3).

Extension to Covert and Malicious Parties. As described in Sect. 2.3.1.2, standard GC protocols secure against covert or malicious adversaries employ cut-and-choose over multiple GCs $\widetilde{f}_i$ derived from seeds s_i.

These protocols similarly benefit from our separation of the server into $\mathcal{S}$ and $\mathcal{T}$. As in the semi-honest protocol, the GC generation can be naturally offloaded to $\mathcal{T}$, achieving corresponding computation and communication relief on the server and network resources. GC correctness verification is achieved by requesting $\mathcal{S}$ to reveal the generator seeds of the circuits to be opened. (Of course, these "opened" circuits are not evaluated.) Note that requirements on $\mathcal{T}$ are the same as in the semi-honest setting. Further, in both covert and malicious cases, the communication between $\mathcal{C}$ and $\mathcal{S}$ is independent of the size of f. The resulting communication complexity of these protocols is summarized in Table 4.2.

Security. We present our protocols implicitly, by describing the modifications to the base protocols of [142]. We informally argue the security of the modifications as they are described. Formal proofs can be naturally built from proofs of [142] and our security arguments. At the very high level, security against $\mathcal{S}/\mathcal{T}$ follows from the underlying GC protocols, since $\mathcal{S}$ is not stronger here than in the two-party SFE setting. The additional power of $\mathcal{C}$ to control the channel between $\mathcal{S}$ and stateful $\mathcal{T}$ is negated by establishing a secure channel (Sect. 4.1.4.3). $\mathcal{C}$'s power to reset stateless $\mathcal{T}$ is addressed by ensuring that by replaying old messages $\mathcal{C}$ gets either what she already knows, or completely unrelated data (Sect. 4.1.4.4).

Table 4.2 Communication complexity

Security	SW-based SFE	Token-assisted SFE
Semi-honest	[231] $\mathcal{O}(|f| + n)$	$\mathcal{O}(n)$
Covert	[103] $\mathcal{O}(|f| + sn + r)$	$\mathcal{O}(sn + r)$
Malicious	[150] $\mathcal{O}(s|f| + s^2 n)$	$\mathcal{O}(s^2 n)$

SW-based versus token-assisted SFE: communication between server S and client C for secure evaluation of function f with $n = |x| + |y|$ inputs, statistical security parameter s, and deterrence probability $1 - 1/r$

Fig. 4.3 Example: circuit representation

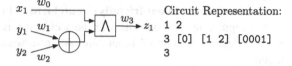

Circuit Representation:
1 2
3 [0] [1 2] [0001]
3

4.1.4.2 Circuit Representation

We now describe our circuit representation format. Our criteria are compactness, the ability to accommodate free XOR gates [142], and the ability of $\mathcal{T}$ to process the encoding "online", i.e., with small constant-size memory. Recall, $\mathcal{T}$ operates in request-response fashion: C incrementally, gate-by-gate, "feeds" the circuit description to $\mathcal{T}$ which responds with the corresponding garbled tables.

We consider circuits with two-input boolean gates. We note that our techniques can be naturally generalized to general circuits. A circuit can efficiently be transformed into our format as described below.

Token-friendly circuit format. Our format for representing the circuit in a token-friendly way is derived from standard circuit representations, such as that used in SFE tools like Fairplay [157] or TASTY (cf. Sect. 5.2). For readability, we describe the format using a simple example circuit shown in Fig. 4.3. This circuit computes $z_1 = x_1 \wedge (y_1 \oplus y_2)$, where x_1 is the input bit of C and y_1, y_2 are two input bits of S. The corresponding circuit representation shown on the right is composed from the description of the inputs, gates, and outputs as follows.

Inputs and wires: The wires w_i of the circuit are labeled with their index $i = \{0, 1, \ldots\}$. The first $u = |x|$ wires are associated with the input of C, the following $v = |y|$ wires are associated with the input of S, and the internal wires are labeled in topological order starting from index $u + v$ (output wires of XOR gates are not labeled, as XOR gates are incorporated into their successor gates as described in the next paragraph). The first line of the circuit description specifies u and v (Fig. 4.3: $u = 1, v = 2$).

Gates are labeled with the index of their outgoing wire; each gate description specifies its input wires. XOR gates do not have gate tables and are omitted from the description. Rather, non-XOR gates, instead of pointing to two input wires, include two input wire *lists*. If the input list contains more than one wire, these wire values are to be XORed to obtain the corresponding gate input. The gate's description concludes with its truth table. In Fig. 4.3, the second line describes the AND gate, which has index 3, and inputs w_0 and $w_1 \oplus w_2$.

Outputs: The circuit description concludes with $|z|$ lines which contain the indices of the output wires (Fig. 4.3: the only ($|z| = 1$) output wire is w_3).

Large XOR sub-circuits. In our token-friendly circuit format, XOR gates with fan-out >1 occur multiple times in the description of their successor gates. In the worst case, this results in a quadratic increase of the circuit description. To avoid this cost, a (non-XOR) identity gate with arbitrary fanout can be inserted after trees of XOR gates.

Conversion into token-friendly circuit format. In order to convert a given circuit with d-input gates into our token-friendly circuit format described above, the gates are first decomposed into 2-input gates and afterwards the XOR gates are grouped together:

Decomposition into 2-input gates. Decomposing the d-input gates of a circuit into multiple 2-input gates can be done in a straight forward way using Shannon's expansion theorem [198] or the Quine–McCluskey algorithm which results in smaller circuits [219]. For small d (e.g., for the very common case of $d = 3$), the optimal replacement can be found via brute-force enumeration of all possibilities (cf. Sect. 3.2.2).

Grouping of XOR gates. The XOR gates can be grouped together as follows: To each input wire and each output wire of a non-XOR gate i we assign the set $\{i\}$. Afterwards we transfer the gates of the circuit in topological order and annotate the output wire of each XOR gate with the following set which is computed from the sets of its input wires S_1, S_2 as $S = S_1 \oplus S_2 := (S_1 \cup S_2) \setminus (S_1 \cap S_2)$. Finally, the remaining non-XOR gates are output in the token format using the sets associated to the input wires which contain those wires which need to be XORed together for the specific input. As merging two sets of size at most $|f|$ entries each can be done in $\mathcal{O}(|f|)$ operations, the overall complexity of this algorithm is in $\mathcal{O}(|f|^2)$.

4.1.4.3 GC Creation with Stateful Token Using Secure Counter

The main idea of our GC generation with small RAM footprint is having $\mathcal{T}$ generate garbled tables "on-the-fly". This is possible, since for each non-XOR gate, the garbled table can be generated given only the garbled values of the gate's input and output wires. In our implementation, we pseudorandomly derive these garbled values from the session key and their wire indices. The rest of this section contains further details.

Session Initialization. SFE proceeds in sessions, where one session is used to securely evaluate a function once. $\mathcal{T}$ has a secure monotonic session counter ctr which is (irreversibly) incremented at the beginning of each session. The session id sid is set to the incremented state of ctr. (We omit the discussion of re-synchronization of ctr between $\mathcal{T}$ and $\mathcal{S}$ which may be needed due to communication and other errors.) Then, the session key is computed by $\mathcal{S}$ and $\mathcal{T}$ as $K = \mathsf{PRF}_k(\mathsf{sid})$ and subsequently used to provide fresh randomness to create the GC.

As required by the construction of Kolesnikov and Schneider [142] (cf. Sect. 2.2.2.3), the two garbled values of the same wire differ by a global offset Δ. This offset is derived from K at session initialization and kept in RAM throughout the session.

Subsequently, the garbled values for wire w_i are derived on-the-fly from K:

$$\widetilde{w}_i^0 = \mathsf{PRF}_K(i), \quad \widetilde{w}_i^1 = \widetilde{w}_i^0 \oplus \Delta. \tag{4.1}$$

Garbled Gates. $\mathcal{T}$ receives the description of the circuit, line by line, in the format described in Sect. 4.1.4.2, and generates and outputs to $\mathcal{C}$ corresponding garbled gates, using only small constant-size memory. $\mathcal{T}$ first verifies that the gate with the same label had not been processed before. (Otherwise, by submitting different gate tables for the same gate, $\mathcal{C}$ may learn the real wire values.) This is achieved with the monotonically increasing processed gate counter gctr, verifying that gate's label glabel > gctr, and setting gctr = glabel. $\mathcal{T}$ then derives and stores the garbled values of the gate's input and output wires according to Eq. (4.1). (For input lists, the wire's garbled value $\widetilde{w}^0$ is computed as the XOR of the garbled values of the listed wires, and $\widetilde{w}^1$ is set to $\widetilde{w}^0 \oplus \Delta$. Note that this requires constant RAM.) Finally, based on these garbled values, the garbled table of the gate is computed and output to $\mathcal{C}$.

Garbled Outputs. Recall, $\mathcal{T}$ must verify circuit correctness by checking mac generated by $\mathcal{S}$. Thus, $\mathcal{T}$ does not release the output decryption tables to $\mathcal{C}$ until after the successful check. At the same time, the check is not complete until the entire circuit had been fed to $\mathcal{T}$. To avoid having $\mathcal{T}$ store the output decryption tables or involving $\mathcal{S}$ here, $\mathcal{T}$ simply encrypts the output tables using a fresh key K', and outputs K' only upon a successful MAC verification.

4.1.4.4 GC Creation with Stateless Token (No Counter)

As discussed above, while non-volatile secure storage (the session counter ctr) is essential in our protocol of Sect. 4.1.4.3, in some cases, it may be desired to avoid its cost. We now discuss the protocol amendments required to maintain security of SFE with the support of a token whose state can be reset by $\mathcal{C}$, e.g., via power interruption.

First, we observe that $\mathcal{S}$ is still able to maintain state, and choose unique session identifiers sid. However, $\mathcal{T}$ can no longer be assured that sid is indeed fresh as claimed by $\mathcal{C}$. Further, $\mathcal{T}$ does not have a source of independent randomness, and thus cannot establish a secure channel with $\mathcal{S}$, e.g., by running a key exchange.

We begin with briefly describing a replay vulnerability of our protocol of Sect. 4.1.4.3, when $\mathcal{T}$ is executed with the same sid. In one session, $\mathcal{C}$ properly executes SFE. In another session, $\mathcal{C}$ runs $\mathcal{T}$ with the same sid, but feeds $\mathcal{T}$ an incorrect circuit, receiving valid garbled tables for each of the gates, generated for the *same* garbled wire values. Now, even though $\mathcal{T}$ will not accept mac and will not decrypt the output wires, $\mathcal{C}$ had already received them in the first execution. It is easy to see that $\mathcal{C}$ "wins".

Our solution is to ensure that C does not benefit from replaying the same session identifier sid twice to T. To achieve this, we require that each garbled value depends on the (hash of the) entire circuit on which this wire depends, as described below. If C replays an old sid with a different circuit, she will not be able to relate the produced garbled table to a previous output of T.

We associate with each wire w_i a (revealed to C) hash value h_i. For input wires, h_i is the empty string. For each other wire w_i, h_i is derived (e.g., via a cryptographic hash function H such as SHA-256 modeled as Random Oracle (RO)) from the description of the gate g_i (which includes its index i, truth table T_i, and list of inputs; cf. Sect. 4.1.4.2) that emits that wire: $h_i = H(\langle gate\_description \rangle)$. The garbled value of wire w_i now depends on its hash value h_i: $\widetilde{w}_i^0 = \mathsf{PRF}_K(h_i)$ and $\widetilde{w}_i^1 = \widetilde{w}_i^0 \oplus \Delta$. Finally, to enable the computation of the garbled tables, C must feed back to T the hashes h_i of the input wires, and receive from T and keep for future use the hash of the output wire. As noted above, C's attempts to feed incorrect values result in the output of garbled tables that are unrelated to previous outputs of T, and thus do not help C.

We note that this construction for stateless tokens can be used together with the Garbled Row Reduction GC technique of [164] to remove one entry from each garbled table (cf. Sect. 2.2.2.3).

4.1.4.5 Hardware Implementation

For the token-assisted GC creation method with stateful tokens described in Sect. 4.1.4.3, an HW architecture for an FPGA has been designed and implemented in VHDL. As described in [128], the HW requirements are moderate and dominated by AES cores to implement PRF and MAC, and an SHA-256 core to implement H. We summarize and compare the performance of the prototype implementation with previous SW-based solutions next.

Overall, the latency of the HW implementation, determined with ModelSim, is #clock_cycles $= 158G_1 + 312G_2 + 154O + 150$, where G_1, G_2 is the number of 1-input gates, respectively 2-input gates, and O is the number of outputs.

Example 6 The HW implementation generates a GC for our optimized 16-bit comparison circuit of Sect. 3.3.3.1 ($G_1 = 0, G_2 = 16, O = 1$) in 5,296 clock cycles ($\approx 80\,\mu s$ with a 50 MHz clock). In SW, this takes roughly 0.5 s on an Intel Core 2 6420 at 2.13 GHz [154].

Example 7 Generating a GC for AES-128 encryption ($G_1 = 12614, G_2 = 11334, O = 128$) takes $5,549,082$ clock cycles (≈ 84 ms with a 50 MHz clock). In SW, this takes approximately 1 s on an Intel Core 2 Duo at 3.0 GHz [180].

We note that the optimization for grouping large XOR sub-circuits described at the end of Sect. 4.1.4.2 dramatically reduces the amount of communication between C and T: When using this optimization, the size of the AES circuit of Example 7 is $|f| = 1.1$ MB and the garbled AES circuit has size $|\widetilde{f}| = 1.1$ MB. Without this optimization, the circuit has no more 1-input gates ($G_1 = 0$) which results in a faster creation ($3,556,070$ clock cycles), evaluation, and smaller size of the GC ($|\widetilde{f}| \approx 0.7\,\text{MB}$); however, as the lists of inputs to be XORed must be repeated for each use, the size of the circuit would be drastically larger ($|f| = 94.5\,\text{MB}$) which might be a bottleneck if the communication between $\mathcal{C}$ and $\mathcal{T}$ is slow.

4.2 One-Time Programs

In this section we investigate the practicality of One-Time Programs (OTPs), a non-interactive version of GC-based SFE protocols which is able to provably protect against arbitrary side-channel leakage.

We propose an efficiency improvement for OTPs with multiple outputs (Sect. 4.2.2.2) and describe a generic architecture for using OTPs in a modular way to protect against arbitrary side-channel attacks in Sect. 4.2.2.3. Further, we present a Hardware (HW) architecture (Sect. 4.2.3.2) and optimizations (Sect. 4.2.3.3) for efficient evaluation of GC/OTP on memory-constrained embedded systems. Implementation results of our architecture and optimizations on a Field-Programmable Gate Array (FPGA) (Sect. 4.2.3.4) show that provable leakage-resilience via OTP comes at a relatively high cost, but its use might still be justified in high-security applications.

See Also. Parts of the following results are based on [129].

4.2.1 Motivation

In the following we summarize side-channels and protection mechanisms (Sect. 4.2.1.1), motivate how HW-assisted SFE can protect against arbitrary leakage (Sect. 4.2.1.2), and give the objectives of this section (Sect. 4.2.1.3).

4.2.1.1 Side-Channels and Protection

For over a decade, we have seen the power and elegance of side-channel attacks on a variety of cryptographic implementations and devices. These attacks refute the assumption of "black-box" execution of cryptographic algorithms, allow the adversary to obtain (unintended) internal state information, such as secret keys, and consequently cause catastrophic failures of the systems. Often the attacks are on a device in the attacker's possession, and exploit physical side-channels such as observing power consumption [139], emitted radiation [2, 87, 182], and even the memory cache [137, 174, 175]. Moreover, even when no computation is performed, stored secrets may be leaked [200] or read out from Random Access Memory (RAM), which is typically not erased at power-off, allowing, e.g., cold-boot attacks [107].

Hence, from the HW perspective, security has been viewed as more than algorithmic soundness in the black-box execution model (see, e.g., [149, 202, 224, 227]).

Today's practical countermeasures typically address known vulnerabilities, and thus target *not all*, but specific classes of side-channel attacks. The desire for a complete solution motivated the recent burst of theoretical research in *leakage-resilient cryptography*, the area that aims to define security models and frameworks that capture leakage aspects of computation or/and memory. Information leakage is typically modeled by allowing the adversary to learn (partial) memory or execution states. The exact information given to the adversary is specified by the (adversarially chosen) leakage function. Then, the assumption on the function (today, usually the bound on the output length) directly translates into a physical assumption on the underlying device and the adversary. Proving security against such an adversary implies security in the real world with the real device, subject to a corresponding assumption [see Pietrzak [179] for a survey on this strand of research]. We note that many of the results of this new line of research (i.e., leakage assumptions and leakage-resilient constructions), although clearly stated, have not yet been evaluated by practitioners and the side-channel community.[5] Further, efficiency is a major concern with today's solutions, since, e.g., embedded systems on an integrated circuit (IC) have very little cost tolerance.[6]

4.2.1.2 SFE in Hardware and Leakage-Resilience

Efficient SFE in an untrusted environment is a longstanding objective of modern cryptography. Informally, the goal of two-party SFE is to let two mutually mistrusting (polynomially-bounded) parties compute an *arbitrary* function on their private inputs without revealing any information about the inputs, beyond the output of the function (cf. Sect. 2.3.1). SFE has a variety of applications, particularly in settings with strong security and privacy demands. Deployment of SFE was very limited and believed expensive until recent improvements in algorithms, code generation, computing platforms and networks.

Because of the execution flow of GC-based SFE (one party can non-interactively evaluate the function once the inputs have been fixed), the security guarantees of SFE are well-suited to prevent *all* side-channel leakage. Indeed, even GC evaluation in the open reveals no information other than the output. Clearly, it is safe to let the adversary see (as it turns out, even to modify) the entire GC evaluation process. The inputs-related stage of GC can also be made non-interactive with appropriate HW such as the (slightly extended) TPM [106]. [101] observed that a very simple HW, called One-Time Memory (OTM), is sufficient, one that, hopefully, can be manufactured tamper-resistant at low cost. They propose to use OTPs, a combination of GC and

---

[5] Indeed, ongoing work of [206] investigates the practical applicability and usability of theoretical leakage models and the constructions proven secure therein.

[6] At the same time, e.g., the size of private circuits in [122] grows quadratically with the number of wire probes by the adversary.

OTMs, for leakage-resilient computation. Indeed, one of our goals is to evaluate today's performance of OTPs in HW.

4.2.1.3 Our Objectives

Practical efficiency of SFE and leakage-resilient computing is critical. Indeed, in most settings, the technology can only be adopted if its cost impact is acceptably low. We pursue the following two objectives.

First, we aim to mark this (practical efficiency) boundary by considering *HW-accelerated* state-of-the-art GC evaluation, and optimizing it for embedded systems. Implementing SFE (at least partially) in HW promises to significantly improve computation speed and reduce power consumption. We evaluate costs, benefits and trade-offs of HW support for GC evaluation.

Second, we use our GC HW-accelerator to implement OTP and evaluate the efficiency of this provably leakage-resilient protection. The envisioned applications for OTP mentioned in [101] are complex functionalities such as one-time proofs, E-cash, or extreme Software protection (with features such as limited number of executions or temporary transfer of cryptographic abilities). However, the exact circuit sizes of these functions, and hence the OTP practicability of these applications, are not yet clear. We make a first step towards estimating the costs of such complex OTP applications by implementing OTP evaluation of the Advanced Encryption Standard (AES) function. We chose AES as it is relatively complex and allows comparison with existing (heuristic) leakage protection. OTP evaluation of AES can be used for sending a small number of messages securely over a completely untrusted platform (e.g., a computer in an Internet café) using a simple tamper-proof HW token (e.g., a USB token) and the same key for encrypting/authenticating multiple messages.

4.2.2 Non-Interactive GCs and One-Time Programs

Using GCs for SFE, although traditionally considered in the interactive setting between server S and client C (cf. Sect. 2.3.1.1), relies on interaction only as much as does the underlying OT protocol. Consequently, as noted in [53], the round complexity and (non-)interactivity features of the GC protocol are exactly those of the underlying OT. In the non-interactive setting, the server who generates the GC is called the sender S and the client who non-interactively evaluates the GC is called the receiver R.

Related Work. The combination of GC with non-interactive OT in the semi-honest setting was proposed in [106]. For a malicious receiver, OTPs were introduced in [101] using minimal HW assumptions. A construction for non-interactive secure computation with information-theoretic security was given in [104] using multiple HW tokens; [72] give a construction with one HW token. [124] consider

non-interactive secure computation in the malicious model with black-box usage of
a pseudorandom generator and a single round of parallel calls to an OT oracle.

We extend and implement OTP evaluation in HW. Our extension is in the computational model with Random Oracles (ROs), secure against a malicious receiver,
and more efficient than the constructions of [101] and [104].

4.2.2.1 Previous Works

In [106] the authors suggested to extend the TPM [216] for implementing non-
interactive OT, resulting in a non-interactive version of Yao's protocol in the semi-
honest setting. Subsequently, OTPs were introduced in [101]. This approach considers malicious receivers and can be viewed simply as Yao's GC protocol, where
the OT function calls are implemented with OTM devices. An OTM T_i is a simple
tamper-proof HW, which allows a single query of one of the two stored garbled
values $\tilde{x}_i^0, \tilde{x}_i^1$ ([101] suggests using a tamper-proof one-time-settable bit b_i which
is set as soon as the OTM is queried).[7] Further, OTM-based GC execution can be
non-interactive, in the sense that the sender $\mathcal{S}$ can send the GC and corresponding
OTMs to the receiver $\mathcal{R}$, who will be able to execute one instance of SFE on any
input of her choice without further interaction with the sender. As also noted in [104],
the evaluated function can be fully hidden by evaluating a universal function instead.
In practice, one would evaluate a garbled *Universal Circuit* that is programmed
to compute the intended function (cf. Sect. 2.3.1.3). Finally, GCs (and hence also
OTPs) are inherently one-time execution objects (generalizable to k-time execution
by repetition).

A subtle issue in this context, noted and addressed in [101], is the following. Previous GC-based solutions were either in the semi-honest model, or used interaction
during protocol execution, which precluded the receiver $\mathcal{R}$ from choosing her inputs
adaptively, based on the given (and even partially evaluated) GC. This possibility of
adaptively chosen inputs results in possible real attacks by a malicious receiver $\mathcal{R}$
in the non-interactive setting.[8] The solution of [101] is to mask (each) output bit z_j
of the function with a random bit m_j, equal to the XOR of (additional) random bits
$m_{i,j}$ contributed by *each* of the input OTMs T_i, i.e., $m_j = m_{1,j} \oplus m_{2,j} \oplus \cdots$ and
$z_j' = z_j \oplus m_j$. The real-world adversary does not learn the output of the function
before it has queried all OTMs with its inputs, which precludes it from adaptively
selecting the input.[9]

---

[7] Indeed, this is one of the simplest functionalities possible, and one that is hopefully easy to
protect against leakage and tampering (we refer the reader to [101] for extended discussion on such
protection).

[8] From the mathematical perspective, the standard proof of security of GC now does not go through,
since the simulator would have to output to $\mathcal{R}$ the simulated GC (i.e., its garbled tables and output
wire decoding) before knowing the inputs of the malicious receiver.

[9] In the proof, the new simulator is able to produce an indistinguishable simulation, since it only
commits to the output values of the simulated GC when the last OTM is queried, the point at which
the simulator knows the inputs of the malicious receiver.

In the following we present an efficiency improvement (Sect. 4.2.2.2), and a generic architecture for leakage-resilient and tamper-proof computation derived from OTP (Sect. 4.2.2.3).

4.2.2.2 Extending One-Time Programs

As mentioned in Sect. 4.2.2.1, the solution in [101] requires each OTM token to additionally store a string of the size of the output. We propose a practical performance improvement to the technique proposed in [101], which is beneficial for OTP evaluation of functions with many output bits. In our solution each OTM token (in addition to the two garbled values) stores a random string r_i of length of the symmetric security parameter t (cf. Sect. 2.1.1.2). Consequently, our construction results in smaller OTMs when the function to be evaluated has more outputs than t. As a trade off, our security proof utilizes ROs, as we do not immediately see how to avoid their use and have OTM size independent of the number of outputs (cf. Sect. 2.1.5 for our discussion on the ROs model). To additionally allow $\mathcal{C}$ to check that the GC evaluation was performed correctly, we make use of the technique of [164, Sect. 2.4] where $\mathcal{S}$ provides for each output wire a one-way image for each of the two possible garbled output values (cf. Sect. 2.3.2).

Our main idea is to insert a "hold off" gate into each output wire W_j which can only be evaluated once *all* input OTMs have been queried, thus preventing malicious receiver $\mathcal{R}$ from choosing her inputs adaptively. It can be implemented by requiring a call to an ideal cipher E (modeled as an RO and instantiated for example with AES) keyed with data from all OTMs.[10] To implement this, we secret-share a random t-bit value $r \in_R \{0, 1\}^t$ over all OTMs, where t is the symmetric security parameter (cf. Sect. 2.1.1.2). That is, OTM T_i additionally stores a share r_i (released to $\mathcal{R}$ with $\tilde{x}_i$ upon the query), where $r = \bigoplus_i r_i$. Receiver $\mathcal{R}$ is able to recover r if and only if she queried all OTMs.

Figure 4.4b depicts our construction: our version of OTMs T_i, in addition to the two OT secrets $\tilde{x}_i^0, \tilde{x}_i^1$ and the tamper-proof bit b_i, contains a random share $r_i \in_R \{0, 1\}^t$ which is released together with $\tilde{x}_i^{x_i}$ once T_i is queried with input bit x_i. The GC is constructed as usual (e.g., as described in Sect. 2.2.2), with the following exception. On each output wire W_j with garbled outputs $\tilde{z}_j^0, \tilde{z}_j^1$, we append a one-input, one-output OT-commit gate G_j, with no garbled table. We set the output wire secrets of G_j to $\hat{z}_j^0 = E_r(\tilde{z}_j^0), \hat{z}_j^1 = E_r(\tilde{z}_j^1)$. To enable $\mathcal{R}$ to compute the wire output non-interactively, GC also specifies that $\hat{z}_j^b$ corresponds to b.

Note that a malicious $\mathcal{R}$ is unable to complete the evaluation of any wire of GC until all the OTMs have been queried, and all her inputs have been specified. Further, she is not able to lie about the result of the computation, since she can only compute one of the two values $\tilde{z}_j^0, \tilde{z}_j^1$. Demonstration of knowledge of $\tilde{z}_j^i$ serves as a proof for the corresponding output value (cf. Sect. 2.3.2).

---

[10] Alternatively, a hash function H can be used which is less efficient [129].

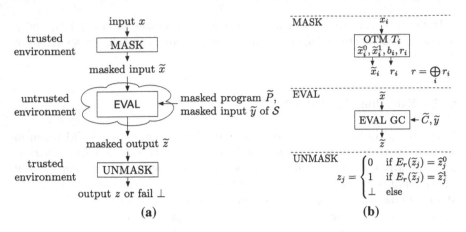

Fig. 4.4 Evaluating a functionality without leakage. **a** Generic architecture. **b** Using One-time memory

Theorem 4 *The above protocol is secure against a semi-honest sender S, who generates the OTM tokens and the GC, and malicious receiver R, in the RO model.*

Proof Security against semi-honest S is trivial as S does not see R's inputs (we consider OTMs a separate entity from S). □

We now describe the simulator Sim which will produce a view indistinguishable from the view of R in real execution. Sim will query the receiver R as a black box and answer all of R's queries, including calls to (simulated) RO O. Our proof is based on the idea that Sim will "program" O such that the output of the "hold off" gates will match the output given by the trusted party of the ideal game.

Without loss of generality, we assume that R queries RO only once for each distinct input. Upon initialization, Sim constructs GC, as would an honest S in the construction described above, and sends the GC to R together with randomly chosen commitments $\widehat{z}_j^0, \widehat{z}_j^1$ for all output wires. Additionally, Sim generates a random key r and a random secret sharing $r = \bigoplus_i r_i$ of it. For the wires corresponding to the input of S, Sim sends secrets corresponding to 0-values. Whenever R queries the ith OTM with input bit x_i, Sim responds with the corresponding garbled value $\widetilde{x}_i^{x_i}$, constructed earlier as part of the GC construction, and the share r_i. Once R has queried the final OTM, Sim sends the input received from R to the trusted party, and receives the output $f(x, y)$ of the computation. Now Sim "programs" O to output certain values according to the received $f(x, y)$. That is, on input $(r, \widetilde{z}_j)$ (call associated with OT-commit gate G_i and the jth bit of the output), O will output $\widehat{z}_j^{f_j(x,y)}$, i.e., the commitment for the wire leaving G_j that corresponds to the bit $f_j(x, y)$ of the output it received from the trusted party.

It is not hard to see that the above simulator generates a view indistinguishable from the view of R in the real execution. First, we note that the simulated GC and responses to RO and OTM queries are indistinguishable from the real execution. Thus, in particular, $\mathcal{R}$ "behaves normally" during the simulation, and would not be able to, e.g., substitute inputs in a special way, etc. Further, "programming" of $\mathcal{O}$ will not be noticed by $\mathcal{R}$, since she can query programmed values only with negligible probability prior to completing all OTM calls (since r is random and unknown to $\mathcal{R}$ prior to completing all OTM calls).

4.2.2.3 Architecture for Using One-Time Programs

Most of today's countermeasures to side-channel attacks are specific to *known* attacks. Protecting HW implementations (e.g., of AES) usually proceeds as follows (e.g., see [4]). First, the inputs are hidden, typically by applying a random mask (this requires trusted operation, and often the corresponding assumption is introduced). Afterwards, the computation is performed on the masked data. To allow this, the functionality needs to be adapted (e.g., using amended AES S-boxes). Finally, the mask is taken off to reveal the output of the computation.

As shown in Fig. 4.4a, we use a similar approach with similar assumptions to provably protect *arbitrary* functionalities against *all* attacks, both known and unknown:

1. The private data x provided by $\mathcal{R}$ is masked in a trusted environment MASK. The masked data $\widetilde{x}$ does not reveal any information about the private data, but still allows computations on it.
2. The computation on the masked data is performed in an untrusted environment where the adversary is able to arbitrarily interfere (passively and actively) with the computation. To compute on the masked data, the evaluation algorithm EVAL needs a specially masked version of the program $\widetilde{P}$. Additionally, $\mathcal{S}$ can provide masked inputs $\widetilde{y}$ to parametrize the program. The result of EVAL is the masked output $\widetilde{z}$.
3. Finally, $\widetilde{z}$ is unmasked into the plain output z. The procedure UNMASK allows verification that $\widetilde{z}$ was computed correctly, i.e., no tampering happened in the EVAL phase in which case UNMASK outputs the failure symbol $\perp$. For correctness of this verification, UNMASK is executed in a trusted environment where the adversary can observe but not modify the computation.

More specifically, the masked program $\widetilde{P}$ is a GC $\widetilde{C}$, masked values $\widetilde{x}, \widetilde{y}, \widetilde{z}$ are garbled values and the algorithms MASK, EVAL and UNMASK can be implemented as described next.

MASK: Masking the input data x of receiver $\mathcal{R}$ is performed by mapping each bit x_i to its corresponding garbled value $\widetilde{x}_i = \widetilde{x}_i^{x_i}$. This can be provided externally (e.g., by interaction with a party on the network). We concentrate on on-board *non-interactive* masking which requires certain HW assumptions (see below). The following can be directly used as a (non-interactive) MASK procedure:

- OTMs [101]: For small functionalities, we favor the very cheap OTMs, as this seems to carry the weakest assumptions (cf. Sect. 4.2). However, as OTMs can be used only once, a fresh OTM must be provided for each evaluation and each input bit. For practical applications, OTMs (together with their GC) could be implemented for example on a tamper-proof USB token for easy distribution.

- Non-interactive OT: TPMs are low-cost tamper-proof cryptographic chips embedded in many of today's PCs [216]. TPM masking based on the non-interactive OT protocol of [106] requires the (slightly extended) TPM to perform asymmetric cryptographic operations in the form of a count-limited private key whose number of usages is restricted by the TPM chip. An interactive protocol allows re-initialization for future non-interactive OT instead of shipping new HW. An alternative approach would be the recently proposed efficient non-interactive OT protocol based on cheap tamper-proof HW tokens that compute symmetric cryptographic operations [73].

- Smartcard: In our preferred solution for larger functionalities, masking could be performed by a tamper-proof smartcard. The smartcard would keep a secure monotonic counter to ensure a single query per input bit. Another advantage of this approach is that the same smartcard can be used to generate the GC as well, thus eliminating GC transfer over the network as described in Sect. 4.1. Further, the smartcard can be naturally used for multiple OTP evaluations.

For non-interactive masking, the HW that masks the inputs must be trusted and the entire input must be specified before anything about the output z is revealed, to prevent adaptive input selection (cf. Sects. 4.2.2.1 and 4.2.2.2).

EVAL: The implementation of EVAL (of the masked program $\widetilde{P}$ on masked inputs $\widetilde{x}$ and $\widetilde{y}$) in embedded systems is presented in detail in Sect. 4.2.3. Here we note that $\widetilde{P}$ and $\widetilde{y}$ (masked input of S) can be generated offline by the semi-honest sender S and provided to EVAL by convenient means (e.g., via a data network or a storage medium). This is the scenario advocated in [101]; one of its advantages is that generation of $\widetilde{P}$ does not leak to EVAL. Alternatively, $\widetilde{P}$ and $\widetilde{y}$ could be generated "on-the-fly" using a cheap, simple, constant-memory, stateless, and tamper-proof token as shown in Sect. 4.1. We reiterate that the masked program $\widetilde{P}$ can be evaluated exactly once.

UNMASK: Finally, the masked output $\widetilde{z}$ is checked for correctness and decoded non-interactively by $\mathcal{R}$ into the plain output z as follows (cf. Sect. 4.2.2.2 and Fig. 4.4b). For each output wire, the masked program $\widetilde{P}$ specifies the correspondence $\widehat{z}_j \rightarrow z_j$ in the form of the two valid encryptions $\widehat{z}_j^0$ and $\widehat{z}_j^1$. Even if EVAL is executed in a completely untrusted environment (e.g., processed on untrusted HW), its correct execution can be verified efficiently: when $E_r(\widetilde{z}_j)$ is neither $\widehat{z}_j^0$ nor $\widehat{z}_j^1$ the garbled output $\widetilde{z}_j$ is invalid and UNMASK outputs the failure symbol $\bot$. The reason for this verifiability property of GC is that a valid garbled output $\widetilde{z}_j$ can only be obtained by correctly evaluating the GC but cannot be guessed as observed in [164, Sect. 2.4] (cf. Sect. 2.3.2).

4.2.2.4 Fully Homomorphic Encryption Against Leakage

At the first glance, the recently proposed fully Homomorphic Encryption (HE) schemes described in Sect. 2.2.1.2 may seem an attractive alternative solution for leakage-free computation. Indeed, fully HE allows computation of arbitrary functions on encrypted data without the need for helper data in the form of a masked program. The MASK algorithm would homomorphically encrypt the input x and the UNMASK algorithm would decrypt the result. Using verifiable computation [91] (cf. Sect. 4.3.3.2), fully HE can also be extended to allow verification that the computation was performed correctly.

However, we argue that fully HE is in fact not appropriate in our setting: our first comment, which concerns any application of fully HE, is that, in its state today, fully HE is extremely computationally intensive as discussed in Sect. 2.2.1.2. Further, even assuming performance similar to that of RSA, this solution would be hundreds of times slower than our GC-based solution, as symmetric primitives used in GC are orders of magnitude faster. Finally, from the leakage-resilience perspective, the UNMASK algorithm will be problematic, as it would need to perform complicated decryptions based on the secret key. We would need to ensure nothing is leaked in these modules, which would bring us either to using much stronger assumptions or to a chicken-and-egg problem.

4.2.3 Evaluating GCs in HW

In this section we describe how GCs (and hence also OTPs) can be efficiently evaluated on embedded systems and memory-constrained devices. First, we discuss how GCs can be evaluated with low memory (Sect. 4.2.3.1). Then, we describe an architecture for evaluating GCs efficiently in HW (Sect. 4.2.3.2) and present compile-time optimizations for that architecture (Sect. 4.2.3.3). Finally, we give timing results on a prototype implementation (Sect. 4.2.3.4).

Running Example: SFE of AES. We stress that our designs and optimizations are generic. However, for concreteness and for meaningful comparison of our HW implementation with prior Software implementations of SFE we choose secure evaluation of the AES functionality as an example. SFE of AES was considered in [180] as a useful and representative function, with applications such as Oblivious Pseudo-random Function (OPRF), blind MACs and encryption, and computation on encrypted data. In this setting, sender S provides AES key k as input y and receiver $\mathcal{R}$ provides a plaintext block m as input x. $\mathcal{R}$ obtains the ciphertext c as output z, where $c = \text{AES}(k, m)$. Recall, during GC evaluation (EVAL), both key and message are masked and hence cannot be leaked.

Fig. 4.5 Circuit of k gates
which requires $\Theta(k)$ memory
for evaluation

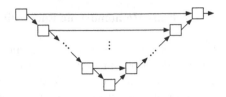

4.2.3.1 GC Evaluation with Low Memory[11]

For evaluating the GC, $\mathcal{C}$ starts with the garbled values of the inputs. Then, $\mathcal{C}$ obtains
the garbled gates one-by-one and evaluates them. The obtained garbled output for
the gate needs to be stored by $\mathcal{C}$ until it is used the last time as input into a garbled
gate.

To reduce the memory footprint to store intermediate values during GC evaluation,
a "good" topologic order can be chosen. As pointed out in [28], the problem of finding
a topologic order of a circuit that minimizes the needed amount of memory to store
intermediate values is equivalent to the register allocation problem. This problem is
well-studied in the context of compilers. In fact, algorithms for register allocation
(e.g., [58, 205]) can be used to find a good topologic order of the circuit which
reduces the amount of memory needed for its evaluation.

In the worst case, the memory needed to evaluate a circuit C is linear in the circuit
size as many intermediate values need to be stored. Figure 4.5 shows an (artificially
constructed) example circuit consisting of k gates which requires $\Theta(k)$ memory for
evaluation.

In our HW architecture presented next, we use only three registers for caching
and leave the investigation of larger cache sizes as future work.

4.2.3.2 Architecture for Evaluating GCs in HW

We describe our architecture for efficient evaluation of GCs on memory-constrained
devices, i.e., having a small amount of slow memory only.

To minimize overhead, we choose key length $t = 127$; with a permutation bit,
garbled values are thus 128 bits long (cf. Sect. 2.2.2). In the following we assume
that memory cells and registers store 128 bit garbled values. This can be mapped to
standard HW architectures by using multiple elements in parallel.

Figure 4.6 shows a conceptual high-level overview of our architecture described
next. At the high-level, EVAL, the process of evaluating a GC, on our architecture
consists of the following steps (cf. Sect. 4.2.2.3). First, the garbled input values $\tilde{x}, \tilde{y}$
are stored in memory using the I/O interface. Then, GC gates are evaluated, using

---

[11] This subsection first appeared in the full version of Järvinen et al. [128] and was published in
Järvinen et al. [127].

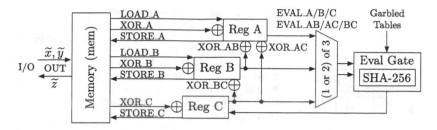

Fig. 4.6 Architecture: GC evaluation on memory-constrained devices

registers A, B, and C to cache the garbled inputs and outputs of a single garbled gate. Finally, the garbled output value $\tilde{z}$ is output over the I/O interface.

As memory access is expensive (cf. Sect. 4.2.3.4) we optimize the code to re-use values already in registers. Our instructions are one-address, i.e., each instruction consists of an operator and up to one memory address. Each of our instructions has length 32 bits: 5 bits to encode one of 18 instructions (described next) and 27 bits to encode an address in memory.

LOAD/STORE: Registers can be loaded from memory using LOAD instructions LOAD_A and LOAD_B. Register C cannot be loaded as it will hold the output of evaluated non-XOR gates (see below). Values in registers can be stored back into memory using STORE instructions STORE_A, STORE_B, and STORE_C.

XOR: We evaluate XOR gates [142] as follows. XOR_A addr computes $A \leftarrow A \oplus \text{mem}[\text{addr}]$. Similarly, the other one-operand XOR operations (XOR1) XOR_B and XOR_C xor the value from memory with the value in the respective register. To compute XOR gates where both inputs are already in registers (XOR2), the instruction XOR_AB computes $A \leftarrow A \oplus B$. Similarly, XOR_AC computes $A \leftarrow A \oplus C$ and XOR_BC computes $B \leftarrow B \oplus C$.

EVAL: Non-XOR gates are evaluated with the Eval Gate block that contains an HW accelerator for SHA-256. As an efficient GC technique we use Garbled Row Reduction [164] as described in Sect. 2.2.2.3. The garbled inputs are in one (EVAL1) or two (EVAL2) registers, and the result is stored in register C. The respective instructions for 1-input gates are EVAL_A, EVAL_B, EVAL_C and for 2-input gates EVAL_AB, EVAL_AC, EVAL_BC. The required garbled table entry for evaluating the garbled non-XOR gate is read from memory.

I/O: The garbled inputs are always stored at the first $|x| + |y|$ memory addresses. The garbled outputs are obtained from memory with OUT instructions.

Example 8 Figure 4.7 shows an example circuit and a possible sequence of instructions to evaluate it on our architecture. First, register A is loaded with $\tilde{x}_1$ from memory address 0×0, then $\tilde{x}_2 \oplus \tilde{y}_1$ is computed in register B and the AND gate is evaluated to obtain output $\tilde{z}_1$ which is stored at address 0×0 and overwrites $\tilde{x}_1$, which is no longer needed. Then, the NOT gate is computed using register B as input and stored at address 0×1. The two outputs $\tilde{z}_1, \tilde{z}_2$ are at addresses 0×0 and 0×1.

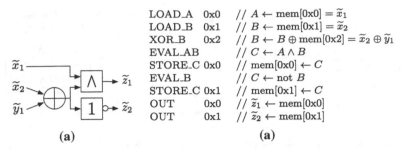

(a)   **(a)**

Fig. 4.7 Example: circuit and instruction sequence (Example 8)

4.2.3.3 Optimizations for Memory-Constrained Devices

In this section, we present several compile-time optimizations to improve performance of GC evaluation (EVAL) on our HW architecture. We aim to reduce the
size of the GC (by minimizing the number of non-XOR gates), the size of the program (number of instructions), the number of memory accesses and memory size
for storing intermediate garbled values. For concreteness, we use AES as a representative functionality for the optimizations and performance measurements, but our
techniques are generic.

Optimization a:Base. Our baseline is the AES circuit/code of [180], already
optimized for a small number of non-XOR gates. Their circuit consists of 11,286
two-input non-XOR gates; thus, its GC has size $11,286 \cdot 3 \cdot 128$ bit ≈ 529 kByte.
Without considering any caching strategies, this results in 113,054 instructions,
hence the program size is $113,054 \cdot 32$ bit ≈ 442 kByte, and the total amount of
memory needed for EVAL is $34,136 \cdot 128$ bit ≈ 533 kByte.

We start with further reduction of the size of the GC.

Optimization b:NoXNOR. First, we reduce the GC size by applying the optimization techniques described in Sect. 3.2.1: We replace XNOR gates with XOR
gates, and propagate the inverted output into the successor gates. Output XNOR
gates are replaced with XOR and a 1-input inverter gate. The cost of this optimization is linear in the size of the circuit [177]. Overall, this optimization results
in the elimination of 4,086 XNOR gates and reduces the size of the AES GC to
$(7,200 \cdot 3 + 40) \cdot 128$ bit ≈ 338 kByte (improvement of 36%).

The remaining optimizations use b:NoXNOR.

Optimization c:Cache. We re-use values already in registers to reduce the number
of LOADs. Values in registers are saved to memory only if needed later.

The remaining optimizations use c:Cache.

Optimization d:MaxFanout. We select a specific topologic order (traversing the
circuit depth-first and following children in decreasing order of their fan-out).

Optimization e:Rand. We randomly consider several orders of evaluation, and
select the most efficient one for EVAL. (This is a one-time compilation expense per
function.) For present work, we considered several random topologic orders of the

Table 4.3 AES circuit optimized for low memory (Sizes in kB)

Optimization	GC $\widetilde{C}$		Program P		Memory for GC evaluation			
	non-XOR	Size	Instr.	Size	Read	Write	Entries	Size
a:Base	11,286	529	113,054	442	67,760	33,880	34,136	533
b:NoXNOR	7,240	338	109,088	426	67,800	33,920	34,176	534
c:Cache	7,240	338	95,885	375	56,779	30,338	21,237	332
d:MaxFanout	7,240	338	74,052	289	42,469	23,767	18,676	292
e:Rand	7,240	338	73,583	287	42,853	22,650	17,315	271

circuit, constructed by the traversal where the next gate is selected at random from the set of unprocessed gates with maximal fan-out. A more rigorous approach to this randomized technique can result in more substantial improvement, and is a possible direction for future work.

Result. Using our optimizations we were able to substantially decrease the memory footprint of EVAL. As shown in Table 4.3, the smallest program was obtained with the non-deterministic optimization e:Rand. This is only slightly better than our best deterministic optimization d:MaxFanout and improves over a:Base as follows. The size of the AES program P is only $73,583 \cdot 32\,\text{bit} \approx 287\,\text{kByte}$ (improvement of 34.9 %). The amount of intermediate memory is $17,315 \cdot 128\,\text{bit} \approx 271\,\text{kByte}$ (improvement of 49.3 %) and the number of memory accesses (read and write) is reduced by $\approx 35\,\%$.

4.2.3.4 Hardware Implementation

We have mapped the HW architecture for evaluating GCs of Sect. 4.2.3.2 to two common system architectures as shown in Fig. 4.8: one for a System on a Programmable Chip (SOPC) with a HW accelerator for a cryptographic hash function (reflecting smartcard and future smartphone architectures) and another for a stand-alone unit (reflecting a custom-made HW accelerator for GC evaluation). In both architectures, the inputs (program P, GC $\widetilde{C}$, and garbled inputs $\widetilde{x}, \widetilde{y}$) and the garbled outputs $\widetilde{z}$ are transferred between the host and the RAM of our HW accelerator over the I/O interface: in the beginning, the host writes the inputs to the RAM and, in the end, the outputs are written to specific addresses from which the host retrieves them.

SOPC. The SOPC architecture shown in Fig. 4.8a consists of a processor (CPU) with access to RAM and a HW accelerator for a cryptographic hash function (e.g., SHA-256) to speed up the most computational burden of the GC evaluation. This is a representative architecture for next generation smartphones or smartcards such as the STMicroelectronics ST33F1M smartcard which includes a 32-bit RISC processor, cryptographic peripherals, and memory comparable to our prototype system [210].

Stand-Alone Unit. The stand-alone architecture shown in Fig. 4.8b consists of a custom-made control state machine, registers (A, B, C), a Hash unit (e.g., SHA-256), and RAM. This architecture could be used to design an Application Specific

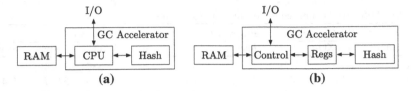

Fig. 4.8 System architectures: hardware-assisted GC evaluation

Table 4.4 Timings:
instructions on prototypes
(clock cycles, average)

Instruction	SOPC	Stand-alone unit
LOAD	291.43	87.63
XOR1	395.30	87.65
XOR2	252.00	1.00
STORE	242.00	27.15
EVAL1	1,282.30	109.95
EVAL2	1,491.68	135.05
OUT	581.48	135.09

IC (ASIC) or an FPGA as a high-speed HW accelerator for GC evaluation. When the
host has written the inputs to the RAM, the stand-alone unit executes the program.
The state machine parses the program and moves data between RAM and registers
or evaluates the non-XOR gates using the Hash unit according to the instructions (cf.
Sect. 4.2.3.2 for details).

Performance Evaluation. Both system architectures (using SHA-256 as cryptographic hash function) have been implemented and evaluated on an Altera/Terasic
DE1 FPGA board with an Altera Cyclone II EP2C20F484C7 FPGA, 512 kB SRAM,
and 8 MB SDRAM running at 50 MHz. For details on the architecture and area
requirements we refer to [129] and concentrate on the timings in the following.

Instruction Timings. The timings of instructions are summarized in Table 4.4.
They show the average number of clock cycles required to execute an instruction
excluding the latency of fetching the instruction. Gate evaluations are expensive in
the SOPC implementation, although the SHA-256 computations are fast, because
they involve a lot of data movement (to/from the Hash unit and from the RAM)
which is expensive. The dominating role of memory reads and writes is clear in the
timings of the stand-alone implementation: the only instructions that do not require
memory operations (XOR2) require only a single clock cycle. Due to the Garbled
Row Reduction GC technique (cf. Sect. 2.2.2.3) which eliminates the first garbled
table entry, EVAL1 is faster than EVAL2 because it accesses the memory on average
every other time compared to three times out of four.

AES Timings. The timings to evaluate the optimized GCs for the AES functionality of Sect. 4.2.3.3 on the prototype implementations are given in Table 4.5. These
timings are for GC evaluation only; i.e, they neglect the time for transferring data
to/from the system because interface timings are highly technology dependent. The
SHA-256 computations take an equal amount of time for both implementations as the

Table 4.5 Timings: GC evaluation of AES on prototypes

| | SOPC | | Stand-alone unit | |
| | Timings (ms) | | Timings (ms) | |
Optimization	SHA	Total	SHA	Total
a:Base	15	1,894	15	225
b:NoXNOR	10	1,749	10	212
c:Cache	10	1,560	10	184
d:MaxFanout	10	1,259	10	144
e:Rand	10	1,253	10	144

Hash unit is the same. The (major) difference in timings is caused by data movement, XORs, interface to the Hash unit, etc. The runtimes for both implementations are dominated by writing and reading the SDRAM; e.g., 84.3 % for the stand-alone unit and our smallest AES circuit (optimization e:Rand). Hence, accelerating memory access, e.g., with burst reads and writes, is the key for further speedups.

Performance Comparison. A Software implementation that securely evaluates the GC/OTP of the unoptimized AES functionality a:Base required 2 s on an Intel Core 2 Duo 3.0 GHz with 4 GB RAM [180]. Our optimized circuit e:Rand evaluated on the stand-alone unit requires only 144 ms for the same operation and, therefore, provides a speedup by a factor of 10.4–17.4 (taking the lack of precision into account).

Clearly, GC/OTP evaluation is substantially slower than evaluation of the plain functionality in HW (a straight forward iterative implementation of AES-128 takes only 10 clock cycles to encrypt one message block). Also the timing overhead when protecting against specific attacks only is substantially smaller (e.g., factor 3.88 for protection against differential power analysis [214]).

Still, using provably leakage-resistant evaluation based on GC/OTP might be justified for applications which require high security guarantees and/or are not evaluated too frequently.

4.3 Application: Privacy-Preserving Cloud Computing

In this section we show how the token for creating GCs of Sect. 4.1.4 can be extended and used for secure outsourcing of data and arbitrary computations thereon in a cloud computing scenario.

Secure outsourcing of computation to an untrusted (cloud) service provider is becoming more and more important. Pure cryptographic solutions based on fully homomorphic and verifiable encryption, recently proposed, are promising but suffer from very high latency. Other proposals perform the whole computation on tamper-proof HW and usually suffer from the same problem.

In the following we focus on applications where the latency of the computation should be minimized, i.e., the time from submitting the query until receiving the result should be as small as possible. To achieve this, we show how to combine a trusted HW token (e.g., a cryptographic coprocessor or a token provided by the customer) with SFE to compute arbitrary functions on secret (encrypted) data where the computation leaks no information and is verifiable. The token is used only in the setup phase whereas in the time-critical online phase the cloud computes the encrypted function on encrypted data using symmetric encryption primitives only and without any interaction with other entities.

See Also. Parts of the following results are based on [190].

4.3.1 Motivation

Enterprises and other organizations often have to store and operate on a huge amount of data. Cloud computing offers infrastructure and computational services on demand for various customers on shared resources. Services that are offered range from infrastructure services such as Amazon EC2 (computation) [8] or S3 (storage) [10], over platform services such as Google App Engine [102] or Microsoft's database service SQL Azure [159], to Software (SW) services such as outsourced customer relationship management applications by Salesforce.com.

While sharing IT infrastructure in cloud computing is cost-efficient and provides more flexibility for the clients, it introduces security risks organizations have to deal with in order to isolate their data from other cloud clients and to fulfill confidentiality and integrity demands. Moreover, since the IT infrastructure is now under control of the cloud provider, the customer has to trust not only the security mechanisms and configuration of the cloud provider, but also the cloud provider itself. When data and computation is outsourced to the cloud, prominent security risks are: malicious code that is running on the cloud infrastructure could manipulate computation and force wrong results or steal data; personnel of the cloud provider could misuse their capabilities and leak data; and vulnerabilities in the shared resources could lead to data leakage or manipulated computation [63]. In general, important requirements of cloud clients are that their data is processed in a confidential way (*confidentiality*), and that their data and computation is processed in the expected way and is not tampered with (*integrity and verifiability*).

Verifiable Computing. Secure outsourcing of *arbitrary* computation and data storage is particularly difficult to fulfill if a cloud client does not trust the cloud provider at all. There are proposals for cryptographic methods which allow specific computations on encrypted data [11], or secure and verifiable outsourcing of storage [133]. Arbitrary computation on confidential data can be achieved by means of verifiable computing. The approach of Gennaro et al. [91] combines fully HE for confidentiality (cf. Sect. 2.2.1.2) with GCs (cf. Sect. 2.2.2) for integrity and verifiability (cf. Sect. 2.3.2). The improved scheme of [62] avoids GCs, but is based on fully HE as well. While these cryptographic schemes can fulfill the aforementioned requirements, they are currently not usable in practice due to the low efficiency of fully HE (cf. Sect. 2.2.1.2).

Trusted Computing. Another line of work tries to solve these problems by establishing trusted execution environments where the cloud client can verify the integrity of the SW and the configuration of the cloud provider's HW platform. This requires, however, secure SW such as secure hypervisors for policy enforcement and attestation mechanisms for integrity verification. The use of trusted computing-based remote attestation in the cloud scenario was recently discussed in [61]. Trusted Virtual Domains [51, 52] are one approach that combines trusted computing, secure hypervisors, and policy enforcement of information flow within and between domains of virtual machines. However, those approaches require trust in a non-negligible amount of HW (e.g., CPU and TPM [216]) which are under the physical control of the cloud provider. According to the specification of the Trusted Computing Group, the TPM is not designed to protect against HW attacks, but provides a shielded location to protect keys. However, the TPM cannot perform arbitrary secure computations on data. It can protect cryptographic keys and perform only pre-defined cryptographic operations like encryption, decryption, and signature creation. In particular, if data should be encrypted it must be provided in plaintext to the TPM, and if data should be decrypted it will be given in plaintext as output. Unfortunately, the TPM cannot be instructed to decrypt data internally, perform computations on the data, and encrypt it again before returning the output. A virtualized TPM [27] that is executed in SW could be enhanced with additional functionality (see, e.g., [188]). However, such SW running on the CPU has access to unencrypted data at some point to compute on it. Hence, if the cloud provider is malicious and uses specifically manipulated HW, confidentiality and verifiability cannot be guaranteed by using trusted computing.

Tamper-Proof Hardware. A HW token which is tamper-proof against physical attacks but can perform arbitrary computations would enable the cloud client to perform confidential and verifiable computation on the cloud provider's site, given that the client trusts the manufacturer of the token that it does not leak any information to the provider. For example, secure coprocessors [203, 204] are tamper-proof active programmable devices that are attached to an untrusted computer in order to perform security-critical operations or to establish a trusted channel through untrusted networks and HW devices to a trusted SW program running inside the secure coprocessor. This can be used to protect sensitive computation from insider attacks at the cloud provider [130]. If cloud providers offer such tokens produced by trustworthy third-party manufacturers, or offer interfaces to attach HW tokens provided by clients to their infrastructure (and by assuming HW is really tamper-proof), cloud clients could perform their sensitive computations inside those tokens. Data can be stored encrypted outside the token in cloud storage while decryption keys are stored in shielded locations of the trusted tokens.

The token-based approach is reasonable because both cryptographic coprocessors and standardized interfaces (e.g., smartcard readers or PCI extension boards) exist that can be used for such tokens. Of course, for trust reasons, the token vendor should not be the same as the cloud provider. However, the whole security-critical computation takes place in the token. Hence, such computation is not really outsourced to the cloud because the function is computed within the token. Some applications, however, require fast replies to queries which cannot be computed online within

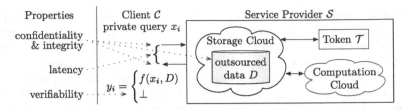

Fig. 4.9 Model: secure outsourcing of data and arbitrary computations

the tamper-proof token. For example, queries in personal health records or payroll databases may not occur very frequently, but need to be processed very fast while privacy of the data should be preserved.

In the following, we focus on cloud application scenarios where private queries to the outsourced data have to be processed and answered with low latency.

Outline and Contribution. The remainder of this section is structured as follows: First, we introduce our model for secure verifiable outsourcing of data and *arbitrary* computations thereon in Sect. 4.3.2. In Sect. 4.3.3 we present architectures to instantiate our model: the first architecture computes the function within a tamper-proof HW token (Sect. 4.3.3.1) and the second architecture is based on fully HE (Sect. 4.3.3.2).

Our main technical contribution is a third architecture (Sect. 4.3.3.3) that combines the advantages of the previous architectures and overcomes their disadvantages. Our solution uses a resource-constrained tamper-proof HW token $\mathcal{T}$ in the setup preprocessing phase. Then, in the online phase only symmetric cryptographic operations are performed in parallel within the cloud without further interaction with $\mathcal{T}$. For this, we adopt the embedded SFE protocol of Sect. 4.1 to the large-scale cloud-computing scenario.

Finally, in Sect. 4.3.4 we compare the performance of all three proposed architectures and show that our scheme allows secure verifiable outsourcing of data and *arbitrary* computations thereon with low latency.

4.3.2 Model for Privacy-Preserving Cloud Computing

We consider the model shown in Fig. 4.9 that allows a client $\mathcal{C}$ to verifiably and securely outsource a database D and computations thereon to an untrusted (cloud) service provider $\mathcal{S}$.

A client $\mathcal{C}$ (e.g., a company) wants to securely outsource data D and computation of a function f (represented as a boolean circuit) thereon to an untrusted service provider $\mathcal{S}$ who offers access to (cloud) storage services and to (cloud) computation services. Example applications include outsourcing of medical data, log files or payrolls and computing arbitrary statistics or searches on the outsourced data. In addition, the evaluation of f can depend on a session-specific private query x_i of $\mathcal{C}$ resulting in the response $y_i = f(x_i, D)$. However, $\mathcal{S}$ should be prevented from learning or modifying D or x_i (*confidentiality and integrity*) and from computing f incorrectly (*verifiability*).[12] Any cheating attempts of a malicious S who tries to deviate from the protocol should be detected by C with overwhelming probability where C outputs the special failure symbol $\perp$.[13]

While this scenario can easily be solved for a restricted class of functions (e.g., private search for a keyword x_i using searchable encryption [133]), we consider the general case of arbitrary functions f. Due to the large size of D (e.g., a database) and/or the computational complexity of f, it is not possible to securely outsource D to S only and let C compute f after retrieving D from S. Instead, the confidentiality and integrity of the outsourced data D has to be protected while at the same time secure computations on D need to be performed at S *without* interaction with C.

4.3.2.1 Tamper-Proof Hardware Token T

To improve the efficiency of the secure computation, our model additionally allows that C uses a *tamper-proof HW token T*, integrated into the infrastructure of S, that is capable of performing computations on behalf of C within a shielded environment, i.e., T must be guaranteed not to leak any information to S. As T needs to be built tamper-proof and cost-effective, it will have a restricted amount of memory only. In many cases the available memory within T will not be sufficient to store D or intermediate values during evaluation of f. If needed, T might resort to additional (slow) secure external memory (e.g., [90]).

The token T could be instantiated with a cryptographic coprocessor built by a third-party manufacturer whom C trusts to ensure that T does not leak any information to S. A possible candidate would be the IBM Cryptographic Coprocessor 4758 or its successor 4764 which is certified under FIPS PUB 140-2 [115, 203]. Such cryptographic coprocessors allow secret keys to be generated internally and securely transported to C or to another token for migration purposes, and authentication to verify that the intended SW is executed within the shielded environment. (For details on migrating a state (key) between two trusted environments (cryptographic coprocessors) we refer to [27, 188]). As such tokens based on cryptographic coprocessors can be used for multiple users in parallel, their costs amortize for service provider and users.

For extremely security critical applications where C does not want to trust the manufacturer of cryptographic coprocessors offered by S, C can choose her own HW manufacturer to produce the tamper-proof HW token T and ship this to S for integration into his infrastructure. We note that this approach is similar to "server hosting" which assumes outsourcing during long periods. However, this contradicts the highly dynamic cloud computing paradigm where service providers can be changed easily.

---

[12] S might attempt to cheat to save storage/computing resources or manipulate the result.

[13] As detailed in Gennaro et al. [91] it must be guaranteed that S cannot learn whether C detected an error or not to avoid that S can use this single bit of information to construct a decryption or verification oracle.

4.3.2.2 Preliminaries and Notation

In the following we make use of fully HE and GCs as defined in Chap. 2.

Notation. $[\![x]\!]$ denotes the fully HE of a value x, and $\tilde{x}$ the garbled value corresponding to x.

Encryption and Authentication. Confidentiality and authenticity of messages can be guaranteed either symmetrically (using one key) or asymmetrically (using two keys).

The symmetric case can be instantiated with a combination of symmetric encryption (e.g., AES [170]) and a MAC (cf. Sect. 2.1.2). These schemes use a respective symmetric key for encryption/authentication and the same key for decryption/verification.

Alternatively, public-key cryptography (e.g., based on RSA or EC-based ElGamal [75]/Digital Signature Algorithm (DSA) [131]) allows usage of separate keys for encryption/authentication and other keys for decryption/verification. This could be used for example to construct an outsourced database to which new entries can be appended by multiple parties without using shared symmetric keys (cf. Fig. 4.9).

Notation. $\hat{x}$ denotes authenticated and $\overline{x}$ encrypted and authenticated data x.

4.3.3 Architectures for Privacy-Preserving Cloud Computing

In the following we present three architectures for instantiating our model of Sect. 4.3.2.

4.3.3.1 Token Computes

A first approach, also used in [130], is to let the token T compute f as shown in Fig. 4.10.

For this, C and T share symmetric keys for encryption and verification. The encrypted and authenticated database $\overline{D}$ and the authenticated function $\hat{f}$ is stored within the storage cloud of service provider S. In the online phase, C sends the encrypted and authenticated query $\overline{x}_i$ to T and the storage cloud provides $\overline{D}$ and $\hat{f}$ one-by-one. T decrypts and verifies these inputs and evaluates $y_i = f(x_i, D)$ using secure external memory to store intermediate values.[14] If T detects any inconsistencies, it continues evaluation substituting the inconsistent value with a random value, and sets y_i to the failure symbol $\bot$. Finally, T sends the authenticated and encrypted response $\overline{y}_i$ back to C who verifies and decrypts $\overline{y}_i$ to obtain the output y_i.

Performance. In this approach, the latency of the online phase, i.e., the time from sending the query x_i to receiving the response y_i, depends on the performance

---

[14] In the worst case, the amount of required external memory can be up to linear in the size of the function as shown in Sect. 4.2.3.1.

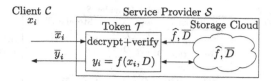

Fig. 4.10 Architecture: token computes [130]

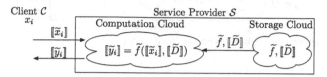

Fig. 4.11 Architecture: cloud computes [91]

of $\mathcal{T}$ (in particular on the performance of secure external memory) and cannot be improved by using the computation cloud services offered by $\mathcal{S}$.

4.3.3.2 Cloud Computes

The approach of [91] shown in Fig. 4.11 does not require a trusted HW token but combines GCS for verifiability (cf. UNMASK step in Sect. 4.2.2.3) and integrity with fully HE for confidentiality of the outsourced data and computations. The main idea is to evaluate the same GC $\tilde{f}$ under fully HE and use the resulting homomorphically encrypted garbled output values to verify that the computation was performed correctly.

Setup Phase. During setup, $\mathcal{C}$ generates a GC $\tilde{f}$ and sends it to $\mathcal{S}$'s storage cloud.[15] To outsource the database D, the corresponding garbled values $\tilde{D}$ are encrypted with the fully HE scheme and $[\![\tilde{D}]\!]$ is stored in $\mathcal{S}$'s storage cloud as well.

Online Phase. In the online phase, $\mathcal{C}$ sends the homomorphically encrypted garbled query $[\![\tilde{x}_i]\!]$ to $\mathcal{S}$ who evaluates the GC $\tilde{f}$ on $[\![\tilde{x}_i]\!]$ and $[\![\tilde{D}]\!]$ using the homomorphic properties of the fully HE scheme. As result, $\mathcal{S}$ obtains $[\![\tilde{y}_i]\!] = [\![\tilde{f}(\tilde{x}_i, \tilde{D})]\!]$ and sends this back to $\mathcal{C}$. After decryption, $\mathcal{C}$ obtains $\tilde{y}_i$ and can verify whether the computation was performed correctly. Otherwise, $\mathcal{C}$ outputs the failure symbol $\perp$.

Improvements. Several improvements have been proposed in Chung et al. [62] to avoid GCs and hence reduce the communication complexity in the setup phase. Still, these constructions require fully HE.

Performance. The advantage of these approaches is that they do not require any trusted HW and hence can be computed in parallel in the computation cloud. However, due to the low performance of today's fully HE schemes, these approaches are unlikely to be used in practical applications in the near future (cf. Sect. 2.2.1.2).

---

[15] This is a correction to the wrong notation and description in Sadeghi et al. [190, Sect. 4.2]

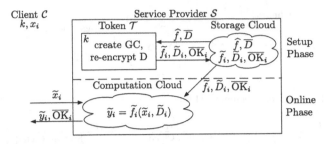

Fig. 4.12 Our architecture: token sets up and cloud computes

Impossibility Results. Indeed, as shown in [222], the setting with a single client is the only one that allows for privacy-preserving cloud computing based on cryptography alone and hence SW only. When data is shared among more than one client, users of cloud services will also need to rely on other forms of privacy enforcement, such as tamper-proof HW, distributed computing, and complex trust ecosystems.

4.3.3.3 Token Sets Up and Cloud Computes

Our approach combines a tamper-proof HW token T used in the setup phase only with efficient computations performed in parallel in the computation cloud as shown in Fig. 4.12. The main idea is that T generates a GC during the setup phase and in the time-critical online phase the GC is evaluated in parallel by the computation cloud.

In detail, our architecture consists of the following three phases:

System Initialization. During *System Initialization*, the client C and the token T agree on a symmetric (long-term) key k (cf. Sect. 4.3.2.1). Additionally, C provides the authenticated function $\widehat{f}$ (represented as a boolean circuit) and the authenticated and encrypted data $\overline{D}$ to S who stores them in the storage cloud.

Setup Phase. In the *Setup Phase*, T generates for protocol invocation i an internal session key k_i derived pseudo-randomly from the key k and i. Using k_i as a seed for randomness generation, T generates a GC $\widetilde{f}_i$ from the function $\widehat{f}$ and a corresponding garbled re-encryption $\widetilde{D}_i$ of the database $\overline{D}$ which are stored in the storage cloud: as described in Sect. 4.1.4.3, the GC can be generated gate-by-gate using a constant amount of memory. For each gate of $\widehat{f}$, S provides T with the description of the gate. T uses the session key k_i to derive the gate's garbled input values and the garbled output value and returns the corresponding garbled table to S. In parallel, T accumulates a hash of the gates requested so far (e.g., by successively updating $h_i = H(h_{i-1} || G_i)$ where H is a cryptographic hash function and G_i is the description of the ith gate) which is finally used to verify authenticity of $\widehat{f}$ (cf. Sect. 4.1.4.1 for details). Similarly, T can convert the authenticated and encrypted database $\overline{D}$ into its garbled equivalent $\widetilde{D}_i$ using constant memory: for each element $\overline{d}$ in $\overline{D}$, T verifies and decrypts $\overline{d}$ and uses the session key k_i to derive the corresponding garbled value $\widetilde{d}_i$ of $\widetilde{D}_i$. Finally, T provides S with an encrypted and authenticated OK message

$\overline{\text{OK}}_i$ that contains the session id and whether the verification of $\hat{f}$ and all elements
in $\overline{D}$ was successful ($\text{OK}_i = \langle i, \top \rangle$) or not ($\text{OK}_i = \langle i, \bot \rangle$).

Online Phase. In the *Online Phase*, C derives the session key k_i and uses this to
create the garbled query $\tilde{x}_i$ which is sent to S. Now, the computation cloud evaluates
the pre-computed GC $\tilde{f}_i$ in parallel using the garbled query and the pre-computed
garbled data $\tilde{D}_i$ as inputs. The resulting garbled output $\tilde{y}_i$ is sent back to C together
with the OK message $\overline{\text{OK}}_i$. Finally, C verifies that both phases have been performed
correctly, i.e., $\overline{\text{OK}}_i$ for the setup phase ($\text{OK}_i = \langle i, \top \rangle$) and valid garbled output keys
$\tilde{y}_i$ for the online phase.

Performance. Our entire architecture is based solely on symmetric cryptographic
primitives and hence is very efficient. When T has access to an HW accelerator for
GC creation [i.e., HW accelerators for Advanced Encryption Standard (AES) and
SHA-256 as described in Sect. 4.1], the performance of the setup phase depends
mostly on the speed of the interface between the token T and the storage cloud
[128]. The size of $\tilde{f}_i$ and $\tilde{D}_i$ is approximately t times larger than the size of $\hat{f}$ and $\overline{D}$,
where t is the symmetric security parameter (cf. Sect. 2.1.1.2). To evaluate the GC
in the online phase, one invocation of SHA-256 is needed for each non-XOR gate
while XOR gates are "free" as described in Sect. 2.2.2.3. GC evaluation can easily be
parallelized for many practical functions that usually perform the same operations
independently on every entry of the database, e.g., computing statistics or complex
search queries.

Extensions. Our architecture can be naturally extended in several ways: To further
speed up the setup phase, a trusted private cloud consisting of multiple tokens can be
used, that in parallel creates GCs and re-encrypts the database for multiple sessions
of the same or different users, or even the same session [50]. The function and the
database can be updated dynamically when an additional monotonic revision number
is used. Such updates can even be performed by multiple clients C_i by using public
key encryption and signatures as described in Sect. 4.3.2.2.

4.3.4 Performance Comparison

We conclude this section with a qualitative performance comparison of the proposed
architectures and leave a prototype implementation for their quantitative performance
comparison as future work.

As summarized in Table 4.6, the asymptotic complexity of the presented architectures is the same: the client C performs work linear in the size of the inputs x_i
and the outputs y_i, the storage cloud stores data linear in the size of the evaluated
function f and the outsourced data D and the computation performed by the token T
respectively the computation cloud is linear in the size of f. Hence, all three schemes
are equally efficient from a complexity-theoretical point of view.

However, the online latency, i.e., the time between C submitting the encrypted
query x_i to the service provider S and obtaining the result y_i, differs substantially in
practice.

Table 4.6 Complexity: architectures for privacy-preserving cloud computing

Architecture	$\mathcal{T}$ computes (Sect. 4.3.3.1)	Cloud computes (Sect. 4.3.3.2)	$\mathcal{T}$ sets up and cloud computes (Sect. 4.3.3.3)
Computation by $\mathcal{C}$	$\mathcal{O}(|x_i| + |y_i|)$	$\mathcal{O}(|x_i| + |y_i|)$	$\mathcal{O}(|x_i| + |y_i|)$
Communication $\mathcal{C} \leftrightarrow \mathcal{S}$	$\mathcal{O}(|x_i| + |y_i|)$	$\mathcal{O}(|x_i| + |y_i|)$	$\mathcal{O}(|x_i| + |y_i|)$
Storage in cloud	$\mathcal{O}(|f| + |D|)$	$\mathcal{O}(|f| + |D|)$	$\mathcal{O}(|f| + |D|)$
Computation by $\mathcal{T}$	$\mathcal{O}(|f|)$ (Online)	None	$\mathcal{O}(|f|)$ (Setup)
Computation by cloud	None	$\mathcal{O}(|f|)$ (Online)	$\mathcal{O}(|f|)$ (Online)
Online latency	$\mathcal{T}$ evals f	Cloud evals $\widetilde{f}(\llbracket \cdot \rrbracket)$	Cloud evals $\widetilde{f}(\cdot)$

For the token-based architecture of Sect. 4.3.3.1, the online latency depends on the performance of the token $\mathcal{T}$ that evaluates f and hence is hard to parallelize and might become a bottleneck in particular when f is large and $\mathcal{T}$ must resort to secure external memory in the storage cloud.

The HE-based architecture of Sect. 4.3.3.2 does not use a token and hence can exploit the parallelism offered by the computation cloud. However, this architecture is not ready for deployment in practical applications yet, as fully HE schemes are not yet sufficiently fast to evaluate a large functionality such as a GC under fully HE.

Our proposed architecture of Sect. 4.3.3.3 achieves low online latency by combining both approaches: $\mathcal{T}$ is used in the setup phase only to generate a GC and to re-encrypt the database. In the online phase, the GC $\widetilde{f}$ is evaluated in parallel by the computation cloud.

Chapter 5
Modular Design of Efficient SFE Protocols

Efficient Secure Function Evaluation (SFE) Techniques. For several years, two approaches for two-party SFE have co-existed—based on either Homomorphic Encryption (HE) (cf. Sect. 2.2.1) or Garbled Circuits (GC) (cf. Sect. 2.3.1). Both approaches have their respective advantages and disadvantages, e.g., GC requires the transfer of the Garbled Circuit (communication complexity is at least linear in the size of the function) but allows almost all expensive operations to be pre-computed resulting in a low latency of the online phase, whereas most HE schemes require relatively expensive public-key operations in the online phase but can result in a smaller overall communication complexity. For a particular primitive, one of the techniques is usually more suitable than the other. For example, for comparisons or computing the maximum, GC is better than HE as described in Sect. 3.4.1, whereas multiplication can benefit from using HE as described in Sect. 5.2.5.1. Therefore, simply switching from one approach for secure computation to the other can result in substantial performance improvements. For instance, for privacy-preserving DNA matching based on secure evaluation of finite automatons, the GC-based protocol of [82] is more efficient than a HE-based one [215].

Combination of Efficient SFE Techniques. Going one step further, it would be beneficial to use the most efficient primitive for the respective sub-task even if they are based on different SFE paradigms. Indeed, secure and efficient composition of sub-protocols based on HE and GC can result in performance improvements as shown for several privacy-preserving applications (e.g., [19, 46, 47, 189]).

5.1 Framework for Modular SFE Protocols

In this section we describe our framework for modularly composing efficient Secure Function Evaluation (SFE) protocols based on Garbled Circuit (GC) and Homomorphic Encryption (HE). We give instantiations of the primitives and conversions of our framework secure in the semi-honest model.

T. Schneider, *Engineering Secure Two-Party Computation Protocols*,
DOI: 10.1007/978-3-642-30042-4_5, © Springer-Verlag Berlin Heidelberg 2012

See Also. Parts of the following results are based on [19, 22, 144] and [145, Sect. IV].

5.1.1 Function Representations

As the complexity of today's most efficient SFE protocols depends linearly on the size of the evaluated function, an obvious approach to improve efficiency is to look for a small representation of the function to be evaluated. However, it is not feasible to describe the optimal choice strategy as finding minimal function representations is hard [33, 132].

Standard representations for functions which are particularly useful for SFE are boolean circuits (cf. Sect. 2.1.3.1) and arithmetic circuits (cf. Sect. 2.1.3.2).

Ordered Binary Decision Diagrams (OBDDs). Another function representation is OBDDs. These can be used to encode decision strategies in a compact way, whereas other functions such as multiplication require exponentially large OBDDs [49, 228]. Protocols for SFE of a function represented as an OBDD are similar to Yao's GC protocol for SFE of boolean circuits, except that they construct and evaluate a garbled OBDD instead of a GC. We do not consider OBDDs explicitly in the following description of our framework, but note that they can be used similarly to GCs. Methods for constructing garbled OBDDs were described first in [147] and improved in [194, Sect. 3.4.1] and [19].

Hybrid Functions. In our framework described below, we advocate a hybrid approach, where function blocks can be represented in any of the above ways, i.e., as boolean circuits, arithmetic circuits, or OBDDs. These blocks are evaluated using the corresponding SFE technique, and their encrypted intermediate results then glued together. This allows choosing the most efficient SFE technique for a specific sub-functionality. For example multiplication of n-bit integer values requires only a single multiplication gate in an arithmetic circuit over a sufficiently large ring, whereas the boolean circuit has size $\mathcal{O}(n^2)$ using the school method or $\mathcal{O}(n^{\log_2 3})$ with the method of Karatsuba and Ofman [134] as described in Sect. 3.3.2. We determine experimentally which method for secure multiplication is the most efficient for a specific input length in Sect. 5.2.5.1.

5.1.2 Modular SFE

In Chap. 2 we have described how arithmetic circuits can be evaluated securely using HE (Sect. 2.2.1.3) and boolean circuits using GCs (Sect. 2.3.1.1). In the following we give a universal framework that combines both approaches and allows conversion back and forth between them. This allows arbitrary compositions of the two techniques and implies significant improvements to SFE.

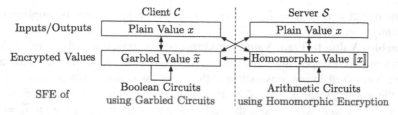

Fig. 5.1 Hybrid SFE protocols

The basic idea of our modular SFE framework is to compose SFE protocols as a sequence[1] of operations on encrypted data as shown in Fig. 5.1: both parties have *Plain Values* as their inputs into the protocol. These plain values, denoted as x, are first encrypted by converting them into their corresponding encrypted value: a *Garbled Value*, denoted as $\tilde{x}$, held by C, or a *Homomorphic Value*, denoted as $[\![x]\!]$, held by S, depending on which operations should be applied. After encryption, the function is securely evaluated on the encrypted values, which may involve conversion of the encryptions into the respective other type of encryption (cf. Sect. 5.1.3). Finally, the encrypted output values are revealed and can be decrypted by converting them into the corresponding plain output values.

In the following we define the two types of encryptions, garbled values (Sect. 5.1.2.1) and homomorphic values (Sect. 5.1.2.2) and describe their conversions from and to plain values. Afterwards we describe methods for converting between the two types of encryptions in Sect. 5.1.3.

5.1.2.1 Garbled Values and Conversions

The interface for GC-based SFE protocols comprises the *garbled values* (cf. Fig. 5.1). As described in Sect. 2.2.2.1, a garbled boolean value $\tilde{x}_i = \langle k_i, \pi_i \rangle$ represents a bit x_i. It consists of a key $k_i \in \{0, 1\}^t$, where t is the symmetric security parameter (cf. Sect. 2.1.1.2), and a permutation bit $\pi_i \in \{0, 1\}$. The garbled value $\tilde{x}_i$ is assigned to one of the two corresponding garbled values $\tilde{x}_i^0 = \langle k_i^0, \pi_i^0 \rangle$ or $\tilde{x}_i^1 = \langle k_i^1, \pi_i^1 \rangle$ with $\pi_i^1 = 1 - \pi_i^0$. A garbled ℓ-bit value can be viewed as a vector of ℓ garbled boolean values.

We show how to convert a plain value into its corresponding garbled value and back next.

Plain Value to Garbled Value for Inputs. To translate a plain value x_i held by S into a garbled value $\tilde{x}_i$ for C, S sends the corresponding garbled value $\tilde{x}_i^0$ or $\tilde{x}_i^1$ to C depending on the value of x_i.

To convert a plain value x_i held by C into a garbled value $\tilde{x}_i$ for C, both parties execute an Oblivious Transfer (OT) protocol where C inputs x_i, S inputs $\tilde{x}_i^0$ and $\tilde{x}_i^1$

---

[1] As all building blocks are secure against semi-honest adversaries, their sequential composition inherits this security property (see e.g., [99]).

and the output to C is $\widetilde{x}_i = \widetilde{x}_i^0$ if $x_i = 0$ or $\widetilde{x}_i^1$ otherwise. OT can be implemented efficiently as described in Sect. 2.2.3.

Garbled Value to Plain Value for Outputs. To convert a garbled value $\widetilde{x}_i = \langle k_i, \pi_i \rangle$ into its corresponding plain value x_i for C, S reveals the output permutation bit π_i^0 which was used during construction of the garbled wire and C obtains $x_i = \pi_i \oplus \pi_i^0$.

If the garbled value $\widetilde{x}_i$ should be converted into a plain value x_i for S, C can simply send $\widetilde{x}_i$ to S who obtains the plain value by decrypting it, e.g., if $\widetilde{x}_i = \widetilde{x}_i^0$, then $x_i = 0$; else $x_i = 1$. We note that malicious C cannot cheat in this conversion as she only knows one of the two garbled values possible and is unlikely to guess the other one. Alternatively, if C is assumed to be semi-honest, it suffices to send π_i to S who obtains $x_i = \pi_i \oplus \pi_i^0$.

5.1.2.2 Homomorphic Values and Conversions

The interface for HE-based SFE protocols are *homomorphic values*, i.e., HEs held by S and encrypted under the public key of C (cf. Fig. 5.1). These homomorphic values can be converted from or to plain values as follows.

Plain Value to Homomorphic Value for Inputs. To convert a plain ℓ-bit value x into a homomorphic value $[\![x]\!]$, x held by S is simply encrypted under C's public key. If x belongs to C, $[\![x]\!]$ is sent to S.

Homomorphic Value to Plain Value for Outputs. To convert a homomorphic value into a plain value for C, S sends the homomorphic value to C who decrypts and obtains the plain value. If only S should learn the plain value corresponding to a homomorphic ℓ-bit value $[\![x]\!]$, S additively blinds the homomorphic value by choosing a random mask $r \in_R \{0, 1\}^{\ell+\sigma}$, where σ is the statistical security parameter (cf. Sect. 2.1.1.2), and computing $[\![\bar{x}]\!] = [\![x]\!] \boxplus [\![r]\!]$. S sends this blinded value to C who decrypts and sends back $\bar{x}$ to S. Finally, S computes $x = \bar{x} - r$.

Packing (cf. Sect. 2.2.1.3) can be used to improve the efficiency of parallel output conversions.

Multiplication of Homomorphic Values with Additively-Homomorphic Encryption. If the HE scheme is only additively homomorphic, two homomorphic values can be multiplied with a single round of interaction as described in Sect. 2.2.1.3. The efficiency of parallel multiplications can be improved by *packing* multiple blinded ciphertexts together.

5.1.3 Conversion Between Encrypted Values

The main idea for converting a homomorphic value $[\![x]\!]$ into a garbled value $\widetilde{x}$ is that addition is a relatively cheap operation under both types of encryptions: One party chooses a random mask and additively blinds the encrypted value. The blinded value is converted into a plain value and encrypted with the target scheme. Finally, the random mask is taken off under encryption.

We give details of the conversion protocols next.

5.1.3.1 Garbled Values to Homomorphic Values

A garbled ℓ-bit value $\widetilde{x}$ held by C (usually obtained from evaluating a GC) can be efficiently converted into a homomorphic value held by S by using additive blinding or bitwise encryption as described next.

Additive Blinding. S randomly chooses a random mask $r \in_R \{0, 1\}^{\ell+\sigma}$, where σ is the statistical security parameter (cf. Sect. 2.1.1.2) and $\ell + \sigma \leq |P|$ to avoid an overflow, and adds the random mask converted into garbled value $\widetilde{r}$ to $\widetilde{x}$ using a garbled $(\ell+\sigma)$-bit addition circuit (cf. Sect. 3.3.1.1) that computes $\widetilde{\overline{x}}$ with $\overline{x} = x + r$. This value is converted into a plain output value $\overline{x}$ for C who homomorphically encrypts this value and sends the result $[\![\overline{x}]\!]$ to S. Finally, S takes off the random mask under encryption as $[\![x]\!] = [\![\overline{x}]\!] \boxplus (-1)[\![r]\!]$. A detailed description of this conversion protocol is given in [144].

Bitwise Encryption. If the bit length ℓ of $\widetilde{x}$ is small, a bitwise approach can be used as well in order to avoid the garbled addition circuit: C homomorphically encrypts the permutation bits π_i of the garbled boolean output values $\widetilde{x}_i = \langle k_i, \pi_i \rangle$ and sends $[\![\pi_i]\!]$ to S. S flips those encrypted permutation bits for which the permutation bit was set as $\pi_i^0 = 1$ during creation to $[\![\pi_i']\!] = [\![1]\!] \boxplus (-1)[\![\pi_i]\!]$ or otherwise sets $[\![\pi_i']\!] = [\![\pi_i]\!]$. Then, S combines these bit encryptions $[\![\pi_i']\!]$ using Horner's scheme as $[\![x]\!] = [\![\pi_\ell' || .. || \pi_1']\!]$.

Performance Comparison. The conversion based on additive blinding requires a garbled addition circuit (cf. Sect. 3.3.1.1) for $(\ell + \sigma)$-bit values and the transfer of the $(\ell + \sigma)$-bit garbled value $\widetilde{r}$. When using Garbled Row Reduction [164] and free XORs [142] as described in Sect. 2.2.2.3, this requires in total $(3 + 1)(\ell + \sigma)(t + 1)$ bits sent from S to C in the pre-computation phase. In the online phase, the GC is evaluated and the result is homomorphically encrypted and sent to S (one ciphertext).

The conversion using bitwise encryption requires ℓ HE and transfer of ℓ ciphertexts from C to S in the online phase. At least for converting a single bit, i.e., when $\ell = 1$, this technique results in better performance.

5.1.3.2 Homomorphic Values to Garbled Values

In the following we describe how to convert a homomorphic ℓ-bit value $[\![x]\!]$ into a garbled value $\widetilde{x}$. This protocol has been widely used to combine HE and GC, e.g., in [19, 46, 47, 126].

S additively blinds $[\![x]\!]$ with a random pad $r \in_R \{0, 1\}^{\ell+\sigma}$, where σ is the statistical security parameter (cf. Sect. 2.1.1.2) and $\ell + \sigma \leq |P|$ to avoid an overflow, as $[\![\overline{x}]\!] = [\![x]\!] \boxplus [\![r]\!]$. S sends the blinded ciphertext $[\![\overline{x}]\!]$ to C who decrypts and inputs the ℓ least significant bits of $\overline{x}$, $\chi = \overline{x} \mod 2^\ell$, to an ℓ-parallel OT protocol (cf. Sect. 2.2.3) to obtain the corresponding garbled value $\widetilde{\chi}$. Then, the mask is taken off within a garbled ℓ-bit subtraction circuit (cf. Sect. 3.3.1.2) which gets as inputs $\widetilde{\chi}$

and $\widetilde{\rho}$ converted from $\rho = r \mod 2^\ell$ as input from $\mathcal{S}$. The output obtained by $\mathcal{C}$ is $\widetilde{x}$ which corresponds to $x = \chi - \rho$.

As proposed in [19], *packing* as described in Sect. 2.2.1.3 can be used to improve parallel conversions from homomorphic to garbled values by packing multiple ciphertexts together before additive blinding and sending them to $\mathcal{C}$.

5.2 Compiling Modular SFE Protocols

In this section we present TASTY, a novel tool for automating, i.e., describing, generating, executing, benchmarking, and comparing, efficient secure two-party computation protocols. TASTY is a new compiler that implements the modular SFE framework described in Sect. 5.1 and can generate protocols based on HE and efficient GCs as well as combinations of both, which often yields the most efficient protocols available today. The user provides a high-level description of the computations to be performed on encrypted data in a domain-specific language. This is automatically transformed into a protocol. TASTY provides most recent techniques and optimizations for practical secure two-party computation with low online latency (cf. Sect. 2.2). Moreover, it allows efficient evaluation of circuits generated by the well-known Fairplay compiler [25, 157].

We use TASTY to compare protocols for secure multiplication based on HE with those based on GCs and fast multiplication (cf. Sect. 3.3.2.2). Further, we show how TASTY improves the online latency for securely evaluating the AES functionality by an order of magnitude compared to previous SW implementations. TASTY allows automatic generation of efficient secure protocols for many privacy-preserving applications where we consider the use case of privacy-preserving face recognition in Sect. 5.3.

See Also. Parts of the following results are based on [109].

5.2.1 Introduction

In the following, we motivate TASTY (Sect. 5.2.1.1), give the outline and our contribution (Sect. 5.2.1.2), and present related works (Sect. 5.2.1.3).

5.2.1.1 Motivation

The design of efficient two-party SFE protocols is vital for a variety of security-critical applications with sophisticated privacy and security requirements such as electronic auctions [164], data mining [152], remote diagnostics [47], classification of medical data [19], and face recognition [76, 173, 189] to name some.

Modern cryptography provides various tools for secure computation. The concept of two-party SFE was introduced in 1982 by Yao [230]. The idea is to let two mutually mistrusting parties compute an arbitrary function (known by both) on their private inputs without revealing any information about their inputs beyond the function's output (cf. Chap. 2 for details). However, the real-world deployment of SFE was believed to be very limited and expensive for a long time. Fortunately, the cost of SFE has been dramatically reduced in recent years thanks to many algorithmic improvements and automatic tools, as well as faster computing platforms and communication networks.

In recent years several cryptographic compilers and specification languages have been proposed that, after a programmer has manually mapped an existing algorithm to integer arithmetic, automatically compile a high-level program into a corresponding SFE protocol. We will give an overview on such previous works in Sect. 5.2.1.3. However, such tools are currently restricted to generating protocols based on only one SFE paradigm, i.e., either GCs or HE, which often results in protocols with suboptimal efficiency. For instance HE allows efficient addition and multiplication of large values (as confirmed by our implementation results in Sect. 5.2.5.1), whereas GCs are better for non-linear functionalities such as comparison (cf. Sect. 3.4.1). As shown in Sect. 5.1, combining the two approaches allows one to obtain relatively efficient protocols when designing privacy-preserving applications, e.g., remote diagnostics [47], classification [19], or face recognition [189].

TASTY is the first compiler that can automatically generate efficient protocols based on HE and GCs as well as combinations of both from a high-level description of the protocol.

5.2.1.2 Outline and Contribution

In summary, TASTY realizes and experimentally verifies many of the concepts and optimizations presented in this book. The remainder of this chapter presents the following contributions in the respective sections.

SFE Compiler. We present TASTY, a tool that allows one to automatically *generate, benchmark* and *compare the performance* of efficient two-party SFE protocols in the semi-honest model (Sect. 5.2.2). We show how TASTY is related to, improves over, and can be combined with existing tools for automatic generation of (two-party) SFE protocols (Sect. 5.2.1.3).

Specification Language. The TASTY Input Language (TASTYL) allows one to *describe* SFE protocols as a *sequence of operations on encrypted data* based on *combinations of GCs and HE*. Hence, it provides a compact and user-friendly description language for describing protocols in the modular SFE framework of Sect. 5.1. TASTYL is based on the Python programming language and *hides technical cryptographic details* from the programmer (Sect. 5.2.3).

Efficient Building Blocks. TASTY implements *efficient building blocks* for HE and GC which allow one to shift most of the complexity into the less time critical setup phase resulting in SFE protocols with a *low-latency online phase* (Sect. 5.2.4). While the implemented techniques were known before, their combination and implementation in a single package is unique and useful. We show how the combination of these techniques speeds up the online phase for *secure evaluation of AES* (a large circuit with more than 30,000 gates) compared to the currently fastest SW implementation of GCs [180] from 5 s to only 0.5 s, while the total costs for setup plus online phase stay almost the same (Sect. 5.2.5.2).

Circuit Optimizations. Additionally, TASTY has built-in tools for *on-the-fly generation and minimization of boolean circuits* (Sect. 5.2.4). As a new circuit building block we implement *fast multiplication circuits* based on Karatsuba's method [134] (cf. Sect. 3.3.2.2). We show that this is more efficient than textbook multiplication (used in previous SFE tools) already for 20-bit numbers.

Benchmarking. Using TASTY, we obtain measurements for a detailed *performance comparison of multiplication protocols based on GCs with those based on HE*. Our experiments show that GC-based multiplication has large communication and time complexity in the setup phase, but results in a more efficient online time than HE-based multiplication for small values (Sect. 5.2.5.1). In particular, multiplication of two garbled values with bit length $\ell \leq 16$ bits requires less online communication and time than the multiplication of two homomorphically encrypted values for short-term security parameters.

Applications. TASTY is a usable and useful tool for describing and automatically generating efficient protocols for several privacy-preserving applications. As a representative example we concentrate on privacy-preserving face recognition in Sect. 5.3. Further application examples are secure set intersection [81, 109, 152] and privacy-preserving medical diagnostics [19, 20, 185, 186].

5.2.1.3 Existing Tools for Two-Party SFE

While the theoretical foundations of two-party SFE were laid already in the 1980s [230, 231], recent optimizations and prototype implementations show that SFE is ready to be used in practical applications (e.g., [154, 180]). To allow the deployment of SFE in a wide range of privacy-preserving applications it is not only important to maximize the *efficiency* of SFE protocols, but also to make SFE *usable* by automatically generating protocols from high-level descriptions. For this, several frameworks for SFE consisting of languages and corresponding tools have been developed in recent years. We review these proposals briefly in the following.

Existing SFE frameworks can be divided into three classes on different abstraction levels as summarized in Table 5.1.

Function Description languages allow specification of *what* function should be computed securely. The function is described in a domain-specific high-level

Table 5.1 Abstraction levels: automatic generation of SFE protocols

Abstraction level	Primitives
Function description	I/O, computation
Protocol description	I/O, enc/dec, computation under encryption
Protocol implementation	I/O, protocols, messages, crypto primitives

programming language which allows programmers to write programs using SFE without any expert knowledge about SFE. Functions described in such languages can then be (formally) analyzed to ensure security of the function (e.g., no information leak to the other party) and are compiled (potentially through lower-level SFE languages) into SFE protocols. Examples are Fairplay's Secure Function Definition Language (SFDL) [25, 157] which can be compiled to boolean circuits (see below), or the Secure Multiparty Computation Language (SMCL) [169] and its Python-based successor PySMCL [167] which allow compilation into arithmetic circuit-based Secure Multi-party Computation (SMPC) protocols such as the Virtual Ideal Functionality Framework (VIFF) [69].

Protocol Description languages allow one to specify *how* the SFE protocol is composed as a sequence of basic operations on encrypted (or secret-shared) data. Examples (described in more detail below) are VIFF [69], the Secure Multiparty Computation language (SMC) [172, 199], Sharemind [32], and the compiler of MacKenzie et al. [155]. These languages allow specification of SFE protocols while abstracting away the details of the underlying cryptographic protocols. The language and compiler we present in this section also fall into this class. However, in contrast to previous work which was restricted to using HE only, our compiler TASTY allows arbitrary combinations of computations under encryption based on GCs and/or HE for *highly efficient* SFE protocols.

Protocol Implementation languages allow one to describe *how exactly* the target SFE protocol is composed as a sequence of basic cryptographic protocol building blocks. They reside at the lowest level of the abstraction hierarchy and require a substantial amount of expert knowledge in cryptographic protocol design. For example the L1 language [195, 196] allows description of secure computation protocols as a sequence of basic primitives such as Oblivious Transfer (OT), encryption/decryption, creation and evaluation of GCs, and messages to be exchanged. Qilin [160] is a Java library for rapid prototyping of cryptographic protocols which currently provides common cryptographic protocols (e.g., OT [163] and coin flipping) using cryptographic primitives (e.g., Pedersen Commitment [178] and ElGamal [75]) implemented with Elliptic Curves (ECs).

Next we describe SFE frameworks which are closely related to ours. In contrast to TASTY, the existing SFE frameworks are based on *either* GCs *or* HE, but not combinations of both.

Garbled Circuits (GCs). The most prominent example of automatic generation of SFE protocols is Fairplay [157] which is based on GCs. Fairplay provides a high-level function description language, SFDL, which allows one to specify the function to be computed securely, i.e., the inputs and outputs of the involved parties, and how the outputs are to be computed from the inputs. The language resembles a simplified version of a HW description language, such as Verilog or Very High Speed Integrated Circuit Hardware Description Language (VHDL). It supports types, variables, functions, boolean operators ($\wedge, \vee, \oplus, \ldots$), arithmetic operators ($+, -$), comparison ($<, \geq, =, \ldots$) and control structures like if-then-else or for-loops with constant range. The Fairplay compiler compiles and optimizes an SFDL program into a boolean circuit which is stored in a file. The circuit can then be evaluated using the Fairplay runtime environment, two Java programs which securely evaluate the circuit using Yao's GC protocol, communicating over a TCP socket. Fairplay is supplemented by FairplayMP [25], a multi-party version of Fairplay suited for three or more parties with the more powerful SFDL 2 input language (with support for $*, /$ and generic functions) and a corresponding circuit compiler. TASTY can serve as an efficient runtime environment for the Fairplay compiler suite, i.e., it allows one to read in circuits generated by the FairplayMP compiler from SFDL 2 programs[2] and optimizes these for efficient secure evaluation with state-of-the-art GC evaluation techniques.

Homomorphic Encryption (HE). VIFF [69], the Virtual Ideal Functionality Framework, is an open source framework written in Python for specifying SMPC protocols as sequences of operations performed on secret-shared (i.e., encrypted) data. While VIFF was mainly designed for secret-sharing based SMPC protocols with three or more parties, it also offers a two-player runtime based on the additively homomorphic Paillier cryptosystem [176]. Using operator overloading, VIFF allows the programmer to express a desired secure computation directly as standard arithmetic without knowing about the used protocol. Indeed, TASTYL, the input language of our compiler, is inspired by the VIFF language, but additionally allows one to combine HE with GC-based computations.

In contrast to general-purpose compilers such as Fairplay, VIFF, and TASTY, the compilers described below are built for specific application scenarios, e.g., use specific number representations [32, 155] or require $n \geq 3$ parties [32, 172, 199]:

The compiler of MacKenzie et al. [155] implements secure two-party computations over values which are secret-shared between the two parties using $\binom{2}{2}$ secret-sharing over a prime field. The computations are composed as a sequence of basic operations on the shared data (e.g., addition or multiplication). The compiler can be used for specific functions such as cryptographic primitives defined over prime fields, e.g., signatures or encryption schemes, where the secret key is shared between both parties.

SMC [172, 199], the Secure Multiparty Computation language, provides a declarative language for describing SMPC based on constraint programming. A program

---

[2] FairplayMP's compiler can also be used to generate circuits for two parties.

is distributed among the parties in the computation along with an interpreter, each party gives its secret inputs and the interpreter calculates the result. Computations are specified as arithmetic circuits and at least 3 parties are required as the underlying multiplication protocol is based on the BGW protocol [26].

Sharemind [32] allows secure computation over the ring of 32-bit integers for three parties and provides an assembly-like programming language. As this setting is fixed and very specific it allows highly efficient protocols.

5.2.2 Tool for Automating Secure Two-Party Computations

In the following we present TASTY, our tool for describing and automatically generating, benchmarking, and evaluating hybrid two-party SFE protocols.

Design Goals. TASTY was designed and developed to meet the following goals:

1. SFE protocols are *programmed* in TASTYL, an intuitive high-level language for describing the protocol as a sequence of operations on encrypted data (cf. Sect. 5.2.3).
2. TASTY allows one to *test, benchmark, and compare* the performance of the generated SFE protocols (cf. [109] for details).
3. The generated SFE protocols aim at *minimizing the latency of the online phase*, i.e., the time from providing the inputs until obtaining the outputs. This is achieved by using a combination of highly efficient primitives and pre-computations (cf. Sect. 5.2.4).

Architecture and Workflow. (cf. Fig. 5.2). The workflow for using TASTY is as follows:

1. Both users, client C and server S, agree on a *Protocol Description* of the SFE protocol in the TASTY Input Language (TASTYL) as described in detail in Sect. 5.2.3.
2. Both users invoke TASTY's *Runtime Environment* (details in [109]), a program that can automatically analyze, run, test, and benchmark the SFE protocol:

   a. In the *Analyzation Phase*, the runtime environment checks the syntactical correctness of the protocol description, exchanges a hash of it to ensure that both parties run the same protocol, and analyzes the protocol to automatically determine which parts of the protocol can be pre-computed.
   b. In the *Setup Phase*, the parties pre-compute those parts of the protocol which are independent of their inputs, e.g., create/send GCs and pre-compute OTs (cf. Sect. 5.2.4 for details).
   c. Finally, in the *Online Phase*, both parties provide their inputs to the computation, and the online part of the SFE protocol is executed (e.g., encryptions and decryptions, online OTs, and evaluation of GCs) to jointly compute the respective outputs for both parties.

Fig. 5.2 Architecture and
workflow of TASTY

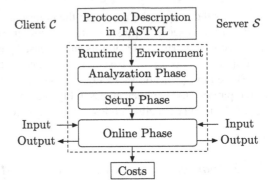

3. TASTY provides a tool to compare the performance costs of multiple SFE protocols (cf. [109] for details).

   Implementation. As implementation language for TASTY we selected Python [181] as it combines elements from both object oriented and functional programming paradigms. In particular the built-in support for generators, a function which yields a value and can be resumed afterwards, was useful for intuitive programming of streamlined large data structures, e.g., for dynamic generation of circuits which allows TASTY to generate and evaluate very large circuits with low memory footprint.

5.2.3 TASTY Input Language

TASTYL, the input language for TASTY, allows one to formulate secure computations as sequence of operations on encrypted data, abstracting away all details of the underlying cryptographic protocols. We start with an overview of the types and operators provided by TASTYL in Sect. 5.2.3.1 and explain the concrete syntax afterwards in Sect. 5.2.3.2.

5.2.3.1 TASTYL Types and Operators

The type system of TASTYL and the operators supported by each type are shown in Fig. 5.3. Each variable in TASTYL is either a scalar *Value* (cf. top half of Fig. 5.3) or a *Vector* (cf. bottom half of Fig. 5.3) which consists of N Values. They can be either unencrypted *Plain Values/Vectors* or encrypted *Garbled or Homomorphic Values/Vectors*.

All Values and Vectors provide the basic operators for (component-wise) addition, subtraction, and multiplication; Vectors also provide dot multiplication: $v \cdot w = \sum_{i=1}^{N} v_i w_i$.

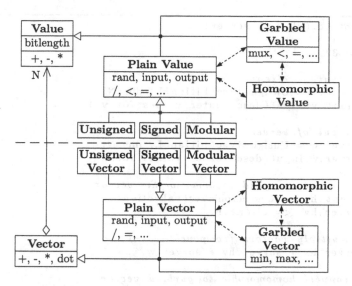

Fig. 5.3 TASTYL: types and operators

Number Representation. Each Value has a bit length ℓ that represents the number of bits needed for its representation. *Unsigned* are unsigned integer values in the range $[0, 2^\ell[$, *Signed* are signed integers in the range $]-2^{\ell-1}, 2^{\ell-1}[$,[3] and *Modular* are elements in the plaintext space of the homomorphic cryptosystem, i.e., $\mathbb{Z}_n$ for Paillier.

In addition to the operations of Value/Vector, the plain/encrypted types support further operations and conversions:

Plain Value/Vector. Inputs and outputs of the two parties are *Plain Values/Vectors*. They can be chosen uniformly at random and provide additional operations (integer) division[4] and comparison.

Homomorphic Value/Vector. Unsigned, Signed and Modular Values/Vectors can be converted into and from homomorphically encrypted *Homomorphic Values/Vectors* of server $\mathcal{S}$. Unsigned and Modular values are mapped directly to $0, \ldots, n-1$. For Signed values, the positive values are mapped to the elements $0, 1, \ldots$ of the plaintext space of the underlying homomorphic cryptosystem, and the negative values to $n-1, n-2, \ldots$ as described in [145]. Addition of two Homomorphic Values/Vectors, and (dot) multiplication of a Homomorphic Value/Vector with a Plain Value/Vector provided by $\mathcal{S}$ is done non-interactively. (Dot) multiplication of two Homomorphic Values/Vectors requires one round of interaction (cf. Sect. 2.2.1.3).

---

[3] Note, we exclude $-2^{\ell-1}$ for signed integers to also allow sign-magnitude representation.

[4] Division by zero or (the unlikely event of) a non-invertible Modular raises an exception.

```
def protocol(client, server):
    N = 4
    L = 32

    # input of client
    client.v = UnsignedVec(bitlen=L, dim=N)
    client.v.input(desc="enter values for v")

    # input of server
    server.w = UnsignedVec(bitlen=L, dim=N)
    server.w.input(desc="enter values for w")

    # convert unsigned to homomorphic vector
    client.hv = HomomorphicVec(val=client.v)
    server.hv <<= client.hv

    # multiply vectors (component-wise)
    server.hx = server.hv * server.w

    # convert homomorphic to garbled vector
    client.gx <<= GarbledVec(val=server.hx)

    # compute minimum value
    client.gmin = client.gx.min_value()

    # convert garbled to unsigned value and output
    client.min = Unsigned(val=client.gmin)
    client.min.output(desc="minimum value")
```

Fig. 5.4 TASTYL: example program

Garbled Value/Vector. Unsigned/Signed Plain and Homomorphic Values/Vectors can be converted into and from *Garbled Values/Vectors* of client C. A Garbled Value can be compared with another one resulting in a Garbled Value of length one bit. This can be used to multiplex (mux), i.e., choose one out of, two Garbled Values. Similarly, the minimum or maximum value and/or index of the components of a Garbled Vector can be determined as Garbled Value(s), e.g., min_value computes the minimum value. For each operation on Garbled Values/Vectors, TASTY automatically infers the underlying GC (cf. Sect. 3.3).

5.2.3.2 TASTYL Syntax and Example

TASTYL is a subset of the Python language; we use the following example to explain its syntax and semantics.

Example 9 (TASTYL Example) Client C and server S have vectors v and w of $N = 4$ unsigned 32-bit values as inputs. As output, C obtains $\min_{i=1,...,N}(v_i \cdot w_i)$. The products $v_i \cdot w_i$ are computed with HE and the minimum with GC.

This protocol can be directly formulated in TASTYL as shown in Fig. 5.4 and described in the following: The protocol gets two parties `client` and `server` as inputs to whom the variables used throughout the protocol are bound (details below). At the beginning, two constants $N = 4$ and $L = 32$ are defined. Then, the input of C, `client.v`, is defined as an unsigned vector of bit length L and dimension N, and read from standard input. Similarly, the input of S, `server.w`, is defined and read. Then, C's input vector `client.v` is converted into a homomorphic vector `server.hv` for S who multiplies this component-wise with his input vector `server.w` resulting in the homomorphic vector `server.hx`. This homomorphic vector is converted into a garbled vector `client.gx` and its minimum value `client.gmin` is computed. Finally, C obtains the intended output by decrypting (converting) `client.gmin` into the unsigned value `client.min`.

Type Conversions. Types can be naturally converted into each other by providing them as input to the constructor of the target type, e.g., in Fig. 5.4, the unsigned vector `client.v` is converted into the homomorphic vector `client.hv` via `client.hv=HomomorphicVec(val=client.v)`. The underlying conversion protocols are described in Sect. 5.1.

Send Operator. The send operator `<<=` transfers variables between the parties, e.g., in Fig. 5.4, hv is sent from C to S with `server.hv <<= client.hv`. When combined with a type conversion, the send operator invokes the corresponding conversion protocol, e.g., in Fig. 5.4, homomorphic vector hx held by S is converted into garbled vector gx held by C with `client.gx <<= GarbledVec(val=server.hx)`.

Binding of Variables. While constants can be declared globally (e.g., N and L in Fig. 5.4), each variable has to be assigned to one of the parties as an attribute.

Inferring Type and Length Automatically. For each operator, TASTY automatically infers the bit length and type of the output variables from those of the input variables s.t. no overflow occurs. Homomorphic values raise an exception if the result does not fit into the plaintext space of the underlying cryptosystem. For example, in Fig. 5.4 the component-wise product of two vectors with N components of unsigned L-bit values results in the homomorphic vector `server.hx` with N components of unsigned 2L-bit values.

Multiple Outputs. GCs can also have multiple garbled output values written as a comma-separated list on the left side of the assignment operator, e.g., the garbled minimum value gv and its index gi can be computed as `(client.gv, client.gi)=client.gx.min\_value\_index()`.

Circuits from File. TASTY allows secure evaluation of boolean circuits read from an external file, e.g., circuits generated by the FairplayMP compiler [25]. For this, the labels of the input and output wires of the circuit are mapped to Garbled Values of corresponding bit length. An example TASTYL file with the concrete syntax for evaluating a garbled file circuit is available at http://tastyproject.net.

5.2.4 Primitives and Optimizations

In TASTY we implemented the following efficient primitives and automatic optimizations that allow one to move expensive operations into the setup phase as precomputations (cf. Fig. 5.2) in order to achieve an online phase with low latency. The modular architecture of TASTY allows extension with other primitives as well. In the following we mention the key features of the used primitives and refer to the previous parts of this book and the original papers for details.

Pre-Defined Security Levels. TASTY has pre-defined security levels following standard recommendations of NIST and ECRYPT II [96] as described in Sect. 2.1.1.2. By using matching basic primitives both security and efficiency are optimized simultaneously. We use EC from the SECG standard [207] and SHA-256 [171] as a cryptographic hash function.

Homomorphic Encryption (HE). We use the additively homomorphic cryptosystem of Paillier [176] (cf. Sect. 2.2.1.1). As key generation for Paillier (an RSA modulus n) is computationally expensive and can be used over multiple protocol runs, the public key is generated and exchanged in the analyzation phase. For efficient encryption we use the extensions of Damgård and Jurik [65, Chap. 6] for precomputing expensive modular exponentiations of the form $r^n \mod n^2$ in the setup phase and only two modular multiplications per encryption in the online phase. As C knows the factorization p, q of n, she uses Chinese remaindering modulo p and q for pre-computing $r^n \mod n^2$ and efficient decryption. Paillier ciphertexts have twice the length of the asymmetric security parameter T as the ciphertext space is $\mathbb{Z}^*_{n^2}$. For modular arithmetics we use gmpy [98], a Python wrapper for the GNU Multiple Precision Arithmetic Library (GMP) [97].

Garbled Circuits (GC). We use the GC construction with free XORs and garbled row reduction [180] secure in the Random Oracle (RO) model (cf. Sect. 2.2.2.3). This GC construction provides free XOR gates (no garbled table and negligible computation). For non-XOR d-input gates, the garbled table consists of 2^{d-1} entries (of size $t + 1$ bits each with symmetric security parameter t); creation requires 2^d and evaluation 1 invocation of SHA-256 modeled as RO.

Circuits. For computations on Garbled Values/Vectors, TASTY dynamically generates circuits using the efficient circuit constructions of Sect. 3.3 which are optimized for a low number of non-XOR gates. Alternatively, circuits can be generated externally, e.g., using the Fairplay compiler [25, 157], and read from a file (cf. Sect. 5.2.3.2). TASTY optimizes the circuits to a low number of non-XOR gates using the optimization of Sect. 3.2.2 which replaces 3-input gates with a low number of 2-input non-XOR gates. XNOR gates are replaced by an XOR gate and an inversion gate which is propagated into successor gates (cf. Sect. 3.2.1.2). Generating, reading, and optimizing circuits is mostly pipelined to allow processing of large circuits with low memory footprint.

Oblivious Transfer (OT). All OT are pre-computed already in the setup phase (cf. Fig. 5.2) using the construction of Beaver [23]; the resulting online phase for OT is highly efficient (transfer and XOR of bitstrings) and depends mostly on the network latency for two messages (cf. Sect. 2.2.3.3). To minimize the computation complexity of the setup phase, we use the efficient OT extension of Ishai et al. [121] to reduce the usually large number of OTs needed in the protocol down to at most t real OTs and some invocations of SHA-256 modeled as RO, where t is the symmetric (computational) security parameter (cf. Sect. 2.2.3.2). The remaining real OT (at most t) are implemented with the OT protocol of Naor and Pinkas [163, Sect. 3.1] using ECs and SHA-256 as RO (cf. Sect. 2.2.3.1). The EC implementation provides point compression to reduce communication at the cost of a negligibly larger computation overhead.

5.2.5 Performance Measurements

We have measured the performance of the primitives implemented in TASTY and compared different protocols against each other and with existing SFE implementations: multiplication protocols based on GC or HE (Sect. 5.2.5.1) and SFE of an AES circuit generated by the Fairplay compiler (Sect. 5.2.5.2).

System Setup. All performance measurements were performed on two desktop PCs with Intel Core 2 Duo CPU (E6850) running at 3.0 GHz and 4 GB Random Access memory (RAM) connected via Gigabit Ethernet. The system runs on 64 bit Gentoo Linux with Python version 2.6.5, gmpy version 1.11 [98] and GMP version 4.3.2 [97]. Unless stated otherwise, all measurements were performed for short-term security (cf. Table 2.2 in Sect. 2.1.1.2) and using point compression for ECs (cf. Sect. 5.2.4).

5.2.5.1 Multiplication Protocols

As arithmetic circuits can express arbitrary computations as a sequence of additions and multiplications (cf. Sect. 2.1.3.2), multiplication is a fundamental basic operation. Indeed, the main difference between SFE protocols based on arithmetic and boolean circuits is the cost of multiplications.

Using TASTY we compare the performance of different secure multiplication protocols based on HE and GCs using fast multiplication circuits (cf. Sect. 3.3.2). For this we constructed four basic test cases. For each SFE paradigm we consider the case where both inputs are provided by one party (S for GC1 and C for HE1), or one by each of the parties (GC2 and HE2). The inputs are Unsigned ℓ-bit values and the output, a 2ℓ-bit Unsigned value, is converted into a Plain output for C. In the following, we compare the communication and the computation complexity of the setup and online phases of the protocols.

Communication. (cf. Fig. 5.5). Our experiments show that GC-based multiplication requires a substantial amount of setup communication (for transfer of GCs) whereas the online communication of GC is better than HE for multiplication of small values. The online communication for multiplying with HE is independent of

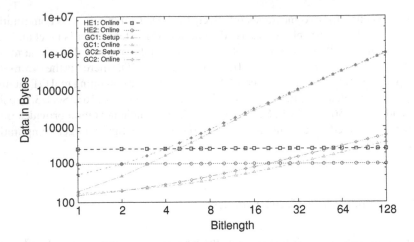

Fig. 5.5 Secure multiplication protocols: communication

the bit length ℓ as a constant number of ciphertexts (2 for HE1 and 5 for HE2) is exchanged. For multiplying with GC, the setup communication grows rapidly due to the large size of the GCs, whereas the online communication complexity grows much more slowly.

Setup Time. (cf. Fig. 5.6a). The time of the setup phase for GC-based multiplication protocols depends on the bit length ℓ as GCs need to be computed; for better visualization we do not plot GC setup times for S in Fig. 5.6a as they are similar to those of C (in our current implementation, C waits until S has created the GC). For HE-based multiplication, the setup time is independent of ℓ as a constant number of encryptions is pre-computed.

Online Time.(cf. Fig. 5.6b). For GC-based multiplication, the time needed by C depends on the size of the evaluated GC which grows with the bit length ℓ; GC's online time for S is negligible. For HE-based multiplication, the time in the online phase is almost independent of ℓ for small bit lengths.

Conclusion. The setup phase for GC-based multiplication is substantially more expensive than that of HE-based multiplication. However, for small values, GC-based multiplication can result in a faster online time than HE-based multiplication. Furthermore, GC-based multiplication, in contrast to HE-based multiplication, needs no (when composed with other GC-based computations) or negligible online interaction and workload for S.

Parallel Multiplications. When N multiplications are done in parallel, e.g., component-wise multiplication of two vectors of N components, time and data complexity of GC-based multiplication grows linearly in N. HE-based parallel multiplication increases more slowly as multiple homomorphic values can be packed before sending from S to C (cf. Sect. 2.2.1.3).

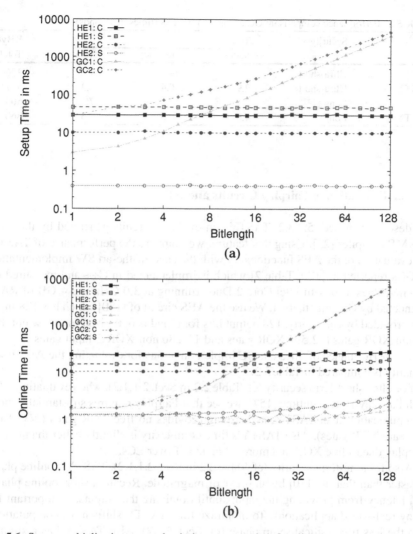

Fig. 5.6 Secure multiplication protocols: timings: **a** setup time, **b** online time

Dependence on Security Level. We note that when the security level is increased to medium- or even long-term security, the performance of HE-based multiplication decreases rapidly while the performance of GC-based multiplication is affected only moderately, as the asymmetric security parameter grows substantially faster than the symmetric one (cf. Table 2.2 in Sect. 2.1.1.2).

Table 5.2 Performance comparison: GC evaluation of AES (times in seconds)

| | Security | Time | | | KByte |
		Setup	Online	Total	Total
[157]	Ultra-short	–	–	4	3760
TASTY	Ultra-short	2.9	0.4	3.3	567
[180]	Long	2	5	7	503
TASTY	Long	4.0	0.5	4.5	860

5.2.5.2 Evaluation of Fairplay Circuits and AES

As described in Sect. 5.2.3.2, TASTY can evaluate circuits generated by the Fair-playMP compiler [25]. Using this feature, we compare the performance of TASTY for evaluation of the AES functionality with the state-of-the-art SW implementation of GCs reported in [180, Table 2] which is implemented in C++ and measured on two machines also with Intel Core 2 Duos running at 3.0 GHz and 4 GB of RAM connected by Gigabit ethernet. We use the AES circuit of [180] which has 128 input bits provided by each party, 128 output bits for C and is optimized for a low number of non-XOR gates (22,594 XOR gates and 11,286 non-XOR 2-input gates).

The performance of different GC implementations for evaluating the AES func-tionality is compared in Table 5.2:

For ultra-short-term security (cf. Table 2.2 in Sect. 2.1.1.2), when evaluating AES with Fairplay's Java runtime [157], we see that Fairplay requires substantially more communication than TASTY, as Fairplay provides no free XOR gates (2/3 of the gates are XOR gates). Also TASTY's time complexity is slightly better than that of Fairplay due to free XOR and more efficient OT over ECs.

Also for long-term security (cf. Table 2.2 in Sect. 2.1.1.2), TASTY's online phase is faster than that of [180] by an order of magnitude. Recall, a short online phase, i.e., latency from providing the inputs until obtaining the outputs, is important for many real-world applications. To minimize this, TASTY shifts most computations into the less time-critical setup phase (cf. Sect. 5.2.2). Also TASTY has a slightly shorter total time than [180], whereas the data complexity is slightly larger due to less optimal data serialization in our implementation. In more detail, the setup time of [180] is 1 s for GC creation and 1 s for data transfer, and the online time is 3 s for OT[5] and 2 s for GC evaluation. In TASTY, the setup time is dominated by 1.1 s for OT and 1.8 s for GC creation, and the online time is dominated by 0.4 s for GC evaluation.

---

[5] As OT seemed not to be the performance bottleneck in Pinkas et al. [180], this implementation used a less efficient, universally composable OT protocol also in the semi-honest setting.

5.3 Application: Privacy-Preserving Face Recognition

Automatic recognition of human faces is becoming increasingly popular in civilian and law enforcement applications that require reliable recognition of humans. However, the rapid improvement and widespread deployment of this technology raises strong concerns regarding the violation of individuals' privacy. A typical application scenario for privacy-preserving face recognition concerns a client who privately searches for a specific face image in the face image database of a server.

In this section we present protocols for privacy-preserving face recognition based on the Eigenfaces algorithm that substantially improve over previous work that is based only on Homomorphic Encryption HE in terms of efficiency and security: the protocol proposed in [76] (Sect. 5.3.3.2) requires $\mathcal{O}(\log M)$ rounds to recognize a face in a database of M faces, whereas the proposal in Barni et al. [21] (Sect. 5.3.3.4) achieves only a weaker definition of security where the client is allowed to learn intermediate information.

Our protocols require only $\mathcal{O}(1)$ rounds and shift most computations into a setup phase. The protocols are based on a combination of HE and GCs in the modular SFE framework of Sect. 5.1. Alternatively, our protocols can also be used for more efficient and secure privacy-preserving fingercode authentication.

See Also. Parts of the following results are based on [109, 189].

5.3.1 Motivation

In the last decade biometric identification and authentication have increasingly gained importance for a variety of enterprise, civilian and law enforcement applications. Examples vary from fingerprinting and iris scanning systems, to voice and face recognition systems. Many governments have already rolled out electronic passports [120] and IDs [162] that contain biometric information (e.g., image, fingerprints, or iris scan) of their legitimate holders.

In particular it seems that facial recognition systems are with increasing frequency installed in surveillance of public places [105], and access and border control at airports [35] to name some. Some of these use cases require online search with short response times and a low amount of communication.

Moreover, face recognition is ubiquitously used also in online photo albums such as Google Picasa and social networking platforms such as Facebook which have become a popular way to share photos with family and friends. These platforms support automatic detection and tagging of faces in uploaded images.[6] Additionally, images can be tagged with the place they were taken.[7]

---

[6] http://picasa.google.com/features-nametags.html; http://face.com

[7] Geotagging can be done either manually or automatically, e.g., on iPhones using GPS http://www.saltpepper.net/geotag.

The widespread use of such face recognition systems, however, raises privacy risks since biometric information can be collected and misused to profile and track individuals against their will. These issues raise the desire to construct privacy-preserving face recognition systems [76].[8]

We concentrate on efficient privacy-preserving face recognition systems. The typical scenario here is a client-server application where the client needs to know whether a specific face image is contained in the database of a server with the following requirements: the client trusts the server to correctly perform the matching algorithm for the face recognition but does not want to reveal any useful information to the server about the requested image or about the outcome of the matching algorithm. The server requires privacy of its database beyond revealing the outcome of the matching algorithm to the client. In [21] it was shown that similar techniques can be used for privacy-preserving fingercode authentication, where a fingerprint is matched against a database of fingerprints (cf. Sect. 5.3.3.4).

In the proposal for privacy-preserving face recognition of Erkin et al. [76] (cf. Sect. 5.3.3.2 for details) the authors use the standard and popular Eigenface [217, 218] recognition algorithm and design a protocol that performs operations on encrypted images by means of HE only. They demonstrate that privacy-preserving face recognition is possible in principle and give required choices of parameter sizes to achieve a good classification rate. However, the proposed protocol requires $O(\log M)$ rounds of online communication as well as computationally expensive operations on homomorphically encrypted data to recognize a face in the database of M faces. Due to these restrictions, the proposed protocol cannot be deployed in practical large-scale applications. We address this aspect and show that one can do better w.r.t. efficiency.

The subsequent proposal of [21] (details in Sect. 5.3.3.4) is also based on HE only and improves efficiency to $O(1)$ rounds at the cost of the client learning additional information.

Our Contribution. We give an efficient and secure privacy-preserving face recognition protocol based on the Eigenfaces recognition algorithm [217, 218] and a hybrid SFE protocol (using the modular SFE framework of Sect. 5.1) which combines the advantages of HE (low communication for linear operations) and GCs (constant round complexity for minimum search). Compared to previous protocols of [76] and [21], our protocol has only a constant number of rounds and shifts most of the computation and communication into a pre-computation phase. The remaining online phase is highly efficient and allows for a quick response time which is especially important in applications such as biometric access control.

Related Work. *Privacy-Preserving Face Recognition* allows a client to obliviously detect if the image of a face is contained in a database of faces held by a server. We give a detailed summary of previous work on privacy-preserving face

---

[8] Similar concerns motivated previous research directions on privacy-preserving iris scanning [44] or fingerprinting [220].

recognition based on the Eigenface recognition algorithm in Sect. 5.3.3.2 for [76] and Sect. 5.3.3.4 for [21].

The subsequent work of [113] considers the simplified setting where parts of the recognition algorithm, namely the Projection phase (cf. Sect. 5.3.2), are not performed securely. As discussed in [76], this setting might be justified in some scenarios where the Eigenfaces are computed from a (sufficiently large) public database of faces, or the server is willing to reveal this information to the client.

Subsequently and independently of our improvements proposed in Sect. 5.3.4, privacy-preserving protocols for iris and fingerprint identification were given in [30]. They show how the techniques underlying our protocols can be improved to better suit their particular setting.

SCiFI, a recently proposed system for secure face identification [173] combines a novel recognition algorithm with a co-designed highly efficient special-purpose SFE protocol. As described in [173], this combination is more accurate and robust (tolerates environmental conditions such as light or persons wearing glasses) than Eigenface-based protocols.

The related problem of *Privacy-Preserving Face Detection* [13] allows a client to detect faces on her image using a private classifier held by a server without revealing the face or the classifier to the other party.

In order to preserve privacy, faces can be de-identified such that face recognition SW cannot reliably recognize de-identified faces, even though many facial details are preserved as described in [165].

5.3.2 Face Recognition Using Eigenfaces

A well-known algorithm for face recognition is the so-called *Eigenfaces* algorithm introduced in [217, 218]. This algorithm achieves reasonable classification rates of approximately 96 % [76] and can be implemented as a privacy-preserving protocol (cf. Sect. 5.3.3). The Eigenfaces algorithm transforms face images into their characteristic feature vectors in a low-dimensional vector space (face space), whose basis consists of *Eigenfaces*. The Eigenfaces are determined through Principal Component Analysis (PCA) from a set of training images; every face is represented as a vector in the face space by projecting the face image onto the subspace spanned by the Eigenfaces. Recognition is done by first projecting the face image into the face space and afterwards locating the closest feature vector. For details on the enrollment process we refer to Erkin et al. [76] and original papers on Eigenfaces [217, 218]. In the following we briefly summarize the recognition process of the Eigenfaces algorithm. A pseudocode description and the naming conventions are given in Algorithm 1; the naming conventions and parameter sizes determined in [76] are listed in Table 5.3.

Inputs and Outputs. The algorithm obtains as input the query face image Γ represented as a pixel image with N pixels. Additionally, the algorithm obtains the parameters determined in the enrollment phase as inputs: the average face Ψ which

Table 5.3 Parameters and sizes: privacy-preserving face recognition

Parameter and Size [76]	Description
M	Number of faces in database
$N = 10304$	Size of a face in pixels
$K = 12$	Number of Eigenfaces
$\Gamma, \Psi \in [0, 2^8 - 1]^N$	Face, average face
$u_1, .., u_K \in [-2^7, 2^7 - 1]^N$	Eigenfaces
$\bar{\Omega}, \Omega_1, .., \Omega_M \in [-2^{31}, 2^{31} - 1]^K$	Projected face, projected faces in database
$D_1, .., D_M \in [0, 2^{50} - 1]$	Squared distances between projected images
$\tau \in [0, 2^{50} - 1]$	Threshold value

is the mean of all training images, the Eigenfaces $u_1, .., u_K$ which span the K-dimensional face space, the projected faces $\Omega_1, .., \Omega_M$ being the projections of the M faces in the database into the face space, and the threshold value τ. The output r of the recognition algorithm is the index of that face in the database which is closest to the query face Γ or the special symbol $\perp$ if no match was found, i.e., all faces have a larger distance than the threshold τ.

Algorithm 1 Face Recognition with Eigenfaces [217, 218]

Input: face Γ, average face Ψ; Eigenfaces $u_1, .., u_K$; projected faces $\Omega_1, .., \Omega_M$; threshold value τ

Output: recognition result $r \in \{1, .., M\} \cup \perp$

    // Phase 1: Projection

1: **for** $i = 1$ to K **do**

2:    $\bar{\omega}_i = u_i^T (\Gamma - \Psi)$

3: **end for**

4: projected face $\bar{\Omega} := (\bar{\omega}_1, .., \bar{\omega}_K)$

    // Phase 2: Distance

5: **for** $i = 1$ to M **do**

6:    compute squared distance $D_i = ||\bar{\Omega} - \Omega_i||^2 = \sum_{j=1}^{K} (\bar{\omega}_j - \omega_{i,j})^2$

7: **end for**

    // Phase 3: Minimum

8: compute minimum value $D_{min} = \min\{D_1, .., D_M\}$ and index i_{min}: $D_{min} = D_{i_{min}}$

9: **if** $D_{min} \leq \tau$ **then**

10:    Return $r = i_{min}$

11: **else**

12:    Return $r = \perp$

13: **end if**

Recognition Algorithm. The recognition algorithm has three phases:

1. Projection: First, the average face Ψ is subtracted from the face Γ and the result is projected into the K-dimensional face space using the Eigenfaces $u_1, .., u_K$. The result is the projected K-dimensional face $\bar{\Omega}$.

2. **Distance**: Afterwards, the square of the Euclidean distance D_i between the projected K-dimensional face $\bar{\Omega}$ and all projected K-dimensional faces in the database Ω_i, $i = 1, .., M$, is computed.
3. **Minimum**: Finally, the minimum distance D_{min} is selected. If D_{min} is smaller than threshold τ, the index of the minimum value, i.e., the identifier i_{min} of the match found, is returned to C as result $r = i_{min}$. Otherwise, no acceptable match was found and the special symbol $r = \perp$ is returned.

5.3.3 Privacy-Preserving Face Recognition

Privacy-Preserving Face Recognition allows a client to obliviously detect if the image of a face is contained in a database of faces held by a server. This can be achieved by securely evaluating a face recognition algorithm within a cryptographic protocol. In the following we concentrate on the Eigenface algorithm described in Sect. 5.3.2 which was also used in [76].

5.3.3.1 Privacy-Preserving Face Recognition Using Eigenfaces

The inputs and outputs of the Eigenfaces algorithm are distributed between client C and server S as shown in Fig. 5.7. Both parties want to hide their inputs from the other party during the protocol run, i.e., C does not want to reveal for which face she is searching while S does not want to reveal the faces in his database or the details of the applied transformation (including Eigenfaces which might reveal critical information about faces in the DB).

In the semi-honest model we are working in (cf. Sect. 2.1.4), parties are assumed to follow the protocol but try to learn additional information from the protocol trace beyond what can be derived from the inputs and outputs of the algorithm when used as a black-box. In particular this requires that all internal results of the Eigenfaces algorithm, including the values passed between the different phases $\bar{\Omega}$ and $D_1, .., D_M$, are "hidden" from both parties. For practical applications it is sufficient to assume that both parties are computationally bounded, i.e., no polynomial-time adversary can derive information from "hidden" values.

We present the initial proposal of [76] for implementing the privacy-preserving Eigenfaces algorithm and "hiding" the intermediate values next.

5.3.3.2 Initial Protocol of [76]

In [76], the authors describe a protocol for privacy-preserving face recognition which implements the Eigenfaces recognition algorithm of Sect. 5.3.2 using HE. Their protocol is secure in the semi-honest model, i.e., players are honest-but-curious [76, Appendix A].

Fig. 5.7 Protocol structure: secure face recognition using eigenfaces

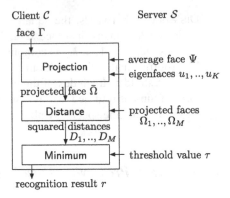

Projection. First, $\mathcal{C}$ and $\mathcal{S}$ jointly compute the projection of the face image Γ into the Eigenspace spanned by the Eigenfaces $u_1, .., u_K$ as follows: $\mathcal{C}$ generates a secret/public key pair of an HE scheme (cf. Sect. 2.2.1.1) and encrypts the face Γ as $[\![\Gamma]\!] = ([\![\Gamma_1]\!], .., [\![\Gamma_N]\!])$. $\mathcal{C}$ sends the encrypted face $[\![\Gamma]\!]$ along with the public key to $\mathcal{S}$. Using the homomorphic properties, $\mathcal{S}$ projects the encrypted face into the low-dimensional face space and obtains the encryption of the projected face $[\![\bar{\Omega}]\!] = ([\![\bar{\omega}_1]\!], .., [\![\bar{\omega}_K]\!])$ by computing for $i = 1, .., K$: $[\![\bar{\omega}_i]\!] = [\![-\sum_{j=1}^{N} u_{i,j}\Psi_j]\!] \boxplus \sum_{j=1}^{N} u_{i,j}[\![\Gamma_j]\!]$. The first factor can already be computed in the pre-computation phase.

Performance Improvement. Additionally we observe that the values $[\![\bar{\omega}_i]\!]$ can be accumulated in parallel by using a parallel fast exponentiation algorithm which re-uses the same squared values of $[\![\Gamma_j]\!]$ in the square-and-multiply method.

Distance. After Projection, $\mathcal{C}$ and $\mathcal{S}$ jointly compute the encryption of the Euclidean distances between the projected face $[\![\bar{\Omega}]\!]$ and all projected faces $\Omega_1, .., \Omega_M$ in the database held by $\mathcal{S}$. This is done by computing for each $i = 1, .., M$: $[\![D_i]\!] = [\![||\Omega_i - \bar{\Omega}||^2]\!] = [\![S_{1,i}]\!] \cdot [\![S_{2,i}]\!] \cdot [\![S_3]\!]$, where $[\![S_{1,i}]\!] = [\![\sum_{j=1}^{K} \omega_{i,j}^2]\!] = \sum_{j=1}^{K}[\![\omega_{i,j}^2]\!]$ and $[\![S_{2,i}]\!] = [\![\sum_{j=1}^{K}(-2\omega_{i,j}\bar{\omega}_j)]\!] = \sum_{j=1}^{K} -2\Omega_{i,j}[\![\bar{\Omega}_j]\!]$ can be computed by $\mathcal{S}$ from $[\![\bar{\Omega}]\!]$ without interaction with $\mathcal{C}$.

We note that the values $[\![S_{1,i}]\!]$ can be pre-computed entirely and online computation of $[\![S_{2,i}]\!]$ can be speeded up by accumulating these values in parallel in order to re-use the same squares in the square-and-multiply exponentiation algorithm.

To obtain $[\![S_3]\!] = [\![\sum_{j=1}^{K} \bar{\omega}_j^2]\!]$ from $[\![\bar{\Omega}]\!]$, the following protocol is suggested in [76]: For $j = 1, .., K$: $\mathcal{S}$ chooses $r_j \in_R \mathbb{Z}_n$, computes $[\![x_j]\!] = [\![\bar{\omega}_j + r_j]\!] = [\![\bar{\omega}_j]\!] \boxplus [\![r_j]\!]$ and sends $[\![x_j]\!]$ to $\mathcal{C}$. $\mathcal{C}$ decrypts $[\![x_j]\!]$, computes $[\![S_3']\!] = [\![\sum_{j=1}^{K} x_j^2]\!]$, and sends $[\![S_3']\!]$ to $\mathcal{S}$. $\mathcal{S}$ finally computes $[\![S_3]\!] = [\![S_3']\!] \boxplus [\![-\sum_{j=1}^{K} r_j^2]\!] \boxplus \sum_{j=1}^{K} -2r_j[\![\bar{\omega}_j]\!]$.

Performance Improvement. As proposed in [189, Full Version, Appendix C], $[\![S_3]\!]$ can be computed more efficiently by choosing shorter random masks and using packing, similar to the parallel multiplication protocol of Sect. 2.2.1.3.

Minimum. Finally, C and S jointly compute the minimum value D from $[\![D_1]\!]$, .., $[\![D_M]\!]$ and its index Id. If the minimum value D is smaller than or equal to the threshold value τ known by S, then C obtains the result Id. To achieve this, [76] suggests the following protocol: Choose the minimum value and index from the list of encrypted value and id pairs $([\![D_0 = \tau]\!], [\![\mathsf{Id}_0 = \bot]\!]), ([\![D_i]\!], [\![\mathsf{Id}_i]\!])_{i=1}^{M}$. For this, they apply a straight forward recursive algorithm for minimum selection based on a sub-protocol which compares two encrypted distances and returns a re-randomized encryption of the minimum and its index to S. For this sub-protocol, they adapt the comparison protocol of Blake and Kolesnikov [29] using the Damgård-Geisler-Krøigaard (DGK) [66–68] cryptosystem (cf. Sect. 3.4.1).

Complexity of Minimum protocol of [76]. The Minimum protocol proposed in [76] requires alogarithmic number of $6\lceil \log_2(M+1)\rceil + 1$ moves. Overall, $8M$ Paillier ciphertexts (of size $2T$ bits each) and $2\ell'M$ DGK ciphertexts (of size T bits each) are sent in the online phase, where $\ell' = 50$ is the length of the squared distances $D_1, .., D_M$ among which the minimum is selected (cf. Table 5.3). This results in a communication complexity of $(16+2\ell')MT$ bits. The asymptotic online computation complexity is dominated by approximately $2M$ Paillier decryptions and $\ell'M$ DGK decryptions for C and the same number of exponentiations for S.

As shown in Sect. 3.4.1, in the two-party case, there exist protocols for computing the minimum of two values that are substantially more efficient than the DGK protocol. These are the key for computing the Minimum step more efficiently and hence improving privacy-preserving face recognition as described in the following.

5.3.3.3 Improved Minimum Protocol of [189]

A first approach is to construct a hybrid protocol using the modular SFE framework of Sect. 5.1, where the Minimum step is computed more efficiently using a GC: For this, the vector of HE $[\![D_1]\!], .., [\![D_M]\!]$ is first converted into its garbled equivalent $\tilde{D}_1, .., \tilde{D}_M$ using the conversion protocol of Sect. 5.1.3.2. Then, a GC is evaluated that computes the minimum value and index as described in Sect. 3.3.3.3. Finally, the resulting minimum distance is compared against the threshold value and, depending on this, either $\bot$ or the minimum index is output to C.

This approach immediately yields constant round complexity and allows one to shift most computations into the setup phase. The resulting complexity of the online phase is substantially more efficient than the protocol of [76] as verified theoretically and experimentally in [189]. Using TASTY, the protocol can be generated automatically from the TASTY Input Language (TASTYL) code shown in Fig. 5.8 [109].

5.3.3.4 Alternative Minimum Protocol of [21]

The techniques for privacy-preserving face recognition can also be adapted for privacy-preserving fingerprint matching as proposed in [21]. In their alternative

```
def protocol(c, s):
  K = 12       # dimension of Eigenspace
  N = 10304    # number of pixels
  M = 42       # size of database

  # Declarations
  s.homegabar = HomomorphicVec(dim=K)
  s.hgamma    = HomomorphicVec(dim=N)
  s.hD        = HomomorphicVec(dim=M)
  c.bot = Unsigned(val=M, bitlen=bitlength(M+1))
  c.gbot = Garbled(val=c.bot)

  # Client inputs
  c.gamma = UnsignedVec(bitlen=8, dim=N).input()

  # Server inputs
  s.omega = UnsignedVec(bitlen=32, dim=(M, K)).input()
  s.psi = UnsignedVec(bitlen=8, dim=N).input()
  s.u = SignedVec(bitlen=8, dim=(K, N)).input()
  s.tau = Unsigned(bitlen=50).input()

  # Projection
  s.hgamma <<= HomomorphicVec(val=c.gamma)
  for i in xrange(K):
    s.homegabar[i] = Homomorphic(val=-(s.u[i].dot(s.psi))) \
      + (s.hgamma.dot(s.u[i]))

  # Distance
  s.hs3 = s.homegabar.dot(s.homegabar)
  for i in xrange(M):
    s.hD[i] = s.hs3 + s.omega[i].dot(s.omega[i])
    s.hD[i] += s.homegabar.dot(s.omega[i]*(-2))

  # Minimum
  c.gD <<= GarbledVec(val=s.hD, force_bitlen=50,
      force_signed=False)
  c.gDmin_val, c.gDmin_ix = c.gD.min_value_index()
  c.gtau <<= Garbled(val=s.tau)
  c.gcmp = c.gDmin_val <= c.gtau
  c.gout = c.gcmp.mux(c.gbot, c.gDmin_ix)
  c.out = Unsigned(val=c.gout)
  if c.out == c.bot:
    c.output("no match found")
  else:
    c.out.output(desc="matched index in DB")
```

Fig. 5.8 TASTYL: improved privacy-preserving face recognition [189]

application scenario, the minimum phase needs to return all the identifiers whose distance is less than a given threshold. Clearly, this is easier to achieve than in the case of privacy-preserving face recognition where this information needs to be hidden from $\mathcal{C}$.

The minimum protocol used in [21] is similar to that of [76] with the following modifications. For better efficiency they use EC-ElGamal instead of DGK encryption (cf. Sect. 3.4.1). To reduce the round complexity from logarithmic to constant, they compare each of the homomorphically encrypted distances $[\![D_i]\!]$ with the threshold value in parallel. We will extend this idea in our further improved hybrid protocol described in Sect. 5.3.4.

Remarks The authors of [21] claim that, in comparison with the Minimum protocols of [76, 189], their protocol as described in [21, Sect. 4.4.1] is

1. functionally equivalent [21, Sect. 4.4.1], and
2. notably more communication efficient [21, Sect. 4.5].[9]

With respect to *functional equivalency* we observe that, in contrast to the protocols of [76, 189], in the protocol of [21, Sect. 4.4.1] the client $\mathcal{C}$ obtains as additional information how many of the entries in the database match the query, i.e., $\#D_i < \tau$.[10] It depends on the application scenario, whether this additional information can be tolerated or not.

With respect to *efficiency* we will show in Sect. 5.3.4 how the protocol of [190] can be slightly changed such that, even when pre-computations are not possible, the protocol is slightly *more* communication efficient than the protocol of [21, Sect. 4.4.1] for the fingerprint matching scenario, while enjoying the additional security property that $\mathcal{C}$ does *not* learn the cardinality of matching entries.

Complexity of Minimum protocol of [21, Sect. 4.4.1]. The Minimum protocol of [21, Sect. 4.4.1] requires a constant number of 5 moves. Using packing, $2M + m'$ Paillier ciphertexts (of size $2T$ bits each) and $2M\ell'$ EC-ElGamal ciphertexts (of size $2(2t + 1)$ bits each using point compression) are sent in the online phase, where m Paillier ciphertexts are needed to transfer M packed $(\sigma + \ell' + 1)$ bit homomorphically encrypted blinded values, where σ is the statistical security parameter (cf. Sect. 2.1.1.2). As $\lfloor \frac{T-1}{\sigma+\ell'+1} \rfloor$ values can be packed into one Paillier ciphertext, $m' = \lceil M / \lfloor \frac{T-1}{\sigma+\ell'+1} \rfloor \rceil$. The resulting communication complexity is $(2M + m') \cdot 2T + 2M\ell' \cdot 2(2t + 1)$ bits. The computation complexity of the protocol is $\mathcal{O}(\ell'M)$ public-key operations for each party.

---

[9] The communication efficiency of the protocol of [21, Sect. 4.4.1] is exactly the same as that of [21, Fig. 2].

[10] Although the outcomes of the comparison do not reveal the identities of the matching entries as they are permuted randomly (see final note in observation [21, Sect. 5.3]), they *do reveal* the cardinality of matching entries.

5.3.4 A Further Improved Hybrid Minimum Protocol

We improve the hybrid Minimum protocol of Sect. 5.3.3.3 by exploiting the observation of Sect. 5.3.3.4 that the distances can be compared first with the threshold and the minimum can be determined afterwards. This results in a smaller circuit and hence better efficiency of the protocol.

The resulting Minimum protocol works as follows: After converting the vector of HE $[\![D_1]\!], .., [\![D_M]\!]$ into its garbled equivalent $\tilde{D}_1, .., \tilde{D}_M$, each garbled value is compared with the threshold value τ using M comparison circuits (cf. Sect. 3.3.3.1). The result is a vector of M garbled bits whose components are encryptions of 1 for each matched entry. Afterwards, the leftmost maximum value and index are determined using a corresponding circuit (cf. Sect. 3.3.3.3) and output to C. If the maximum value is 0, then no match was found. Otherwise, the maximum index is the identifier of the matched entry.

The TASTYL code for the privacy-preserving face recognition protocol using the improved Minimum protocol is listed in Fig. 5.9.

5.3.4.1 Protocol Complexity

We determine the complexity of our further improved Minimum protocol described above: In addition to the M comparison circuits of size ℓ' non-XOR gates each (cf. Sect. 3.3.3.1) the circuit for computing the maximum value and index of 1-bit values requires $2M$ non-XOR gates (cf. Sect. 3.3.3.3). This results in $M(\ell' + 2)$ non-XOR gates.

The M homomorphically encrypted ℓ'-bit values can be converted into their garbled equivalents with the protocol of Sect. 5.1.3.2. This protocol requires evaluation of a circuit with $M\ell'$ non-XOR gates, sends $M\ell'$ garbled bits from S to C, and executes an $\ell'M$ parallel Oblivious Transfer (OT) protocol (cf. Sect. 2.2.3). Additionally, this protocol sends m packed Paillier ciphertexts, where $m = \lceil M / \lfloor \frac{T-1-\sigma}{\ell'} \rfloor \rceil$ as each packed ciphertext can contain up to $\lfloor \frac{T-1-\sigma}{\ell'} \rfloor$ values. Using the OT protocol of Sect. 2.2.3.4 and garbled row reduction [164] with free XORs [142] for GC (cf. Sect. 2.2.2.3) we obtain the following communication complexity in the RO model:

With Pre-Computations. When pre-computations are allowed, most of the communication complexity can be shifted into the setup phase: OTs ($\approx 4M\ell't + 6t^2$ bits), transfer of the GC ($3(t + 1) \cdot 2M(\ell' + 1)$ bits), and transfer of garbled values ($(t + 1)M\ell'$ bits). The resulting communication complexity of the setup phase is $\approx 6t^2 + 11M\ell't$ bits. In the highly efficient online phase, only m Paillier ciphertexts are sent ($2mT$ bits), and the online phase of the OT protocol is executed ($2M\ell't$ bits).

Without Pre-Computations. In ad hoc application scenarios, where C and S meet spontaneously, no pre-computations are possible. In this case the communication complexity of our Minimum protocol consists of the above setup and online communication complexity without the online OT protocol.

```
def protocol(c, s):
  K = 12        # dimension of Eigenspace
  N = 10304     # number of pixels
  M = 42        # size of database

  # Declarations
  s.homegabar = HomomorphicVec(dim=K)
  s.hgamma    = HomomorphicVec(dim=N)
  s.hD        = HomomorphicVec(dim=M)
  c.gC        = GarbledVec(bitlen=1, dim=M)

  # Client inputs
  c.gamma = UnsignedVec(bitlen=8, dim=N).input()

  # Server inputs
  s.omega = UnsignedVec(bitlen=32, dim=(M, K)).input()
  s.psi = UnsignedVec(bitlen=8, dim=N).input()
  s.u = SignedVec(bitlen=8, dim=(K, N)).input()
  s.tau = Unsigned(bitlen=50).input()

  # Projection
  s.hgamma <<= HomomorphicVec(val=c.gamma)
  for i in xrange(K):
    s.homegabar[i] = Homomorphic(val=-(s.u[i].dot(s.psi))) \
        + (s.hgamma.dot(s.u[i]))

  # Distance
  s.hs3 = s.homegabar.dot(s.homegabar)
  for i in xrange(M):
    s.hD[i] = s.hs3 + s.omega[i].dot(s.omega[i])
    s.hD[i] += s.homegabar.dot(s.omega[i]*(-2))

  # Minimum
  c.gD <<= GarbledVec(val=s.hD, force_bitlen=50,
      force_signed=False)
  c.gtau <<= Garbled(val=s.tau)
  for i in xrange(M):
    c.gC[i] = (c.gD[i] <= c.gtau)
  c.gDmax_val, c.gDmax_ix = c.gC.max_value_index()
  c.Dmax_val = Unsigned(val=c.gDmax_val)
  c.Dmax_ix = Unsigned(val=c.gDmax_ix)
  if c.Dmax_val == 0:
    c.output("no match found")
  else:
    c.Dmax_ix.output(desc="matched index in DB")
```

Fig. 5.9 TASTYL: further improved privacy-preserving face recognition

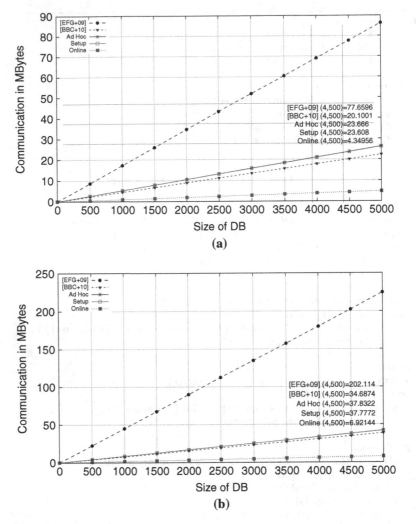

Fig. 5.10 Minimum: communication complexity (face-recognition): **a** ultra-short-term security, **b** long-term security

5.3.4.2 Performance Comparison

Finally, we compare the Minimum protocols presented above with respect to their communication and computation complexity.

Communication Complexity. We choose the size of security parameters according to Sect. 2.1.1.2 and set the statistical security parameter to $\sigma = 80$. For the bit

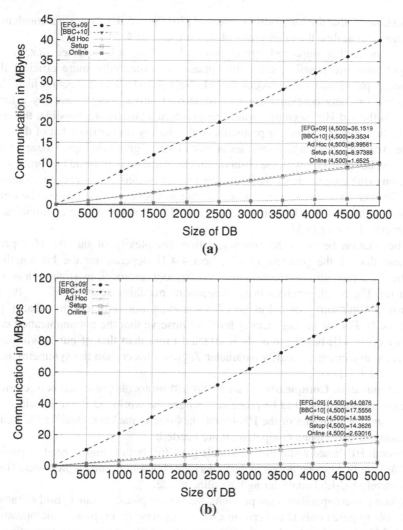

Fig. 5.11 Minimum: communication complexity (fingerprint matching): **a** ultra-short-term security, **b** long-term security

length ℓ' of the values D_i we consider two application scenarios: Eigenface-based face recognition of [76] ($\ell' = 50$) as shown in Fig. 5.10 on p. 119 and Fingerprint matching of [21] ($\ell' = 19$) as shown in Fig. 5.11 on p. 120. In particular we compare the values for a database with $M = 4500$ entries.

We observe that the Minimum protocol of [76] (Sect. 5.3.3.2) requires much more communication than the other protocols (by a factor of 4–5).

For our further improved hybrid protocol of Sect. 5.3.4 we see that, if pre-computations are possible, the online phase is substantially more efficient than the setup phase and the protocol of [21, Sect. 4.4.1] (Sect. 5.3.3.4) which does not allow one any communication to be moved into a setup phase (by a factor of 5). In the Ad Hoc scenario, where no pre-computations are possible, the communication complexity of our protocol is only slightly higher than that of the setup phase in case of pre-computations (see almost identical graphs in Figs. 5.10 and 5.11). This matches perfectly with the observation made above, that the ad hoc protocol, in addition to the communication of the setup phase, transfers only a very small number of m Paillier ciphertexts, where $m \ll M$ due to packing. As expected, the online phase is dominated by the online phase of the OT protocol whose communication complexity is linear in M.

The relation between the communication complexity of our Ad Hoc protocol and that of the protocol of [21, Sect. 4.4.1] depends on the bit length ℓ': in the face recognition scenario ($\ell' = 50$) our protocol is slightly less efficient (cf. Fig. 5.10), whereas in the fingerprint matching scenario ($\ell' = 19$) our improved Minimum protocol of Sect. 5.3.4 is even more efficient than that of [21, Sect. 4.4.1]. For increasing security level we observe that the communication complexity of the HE-based protocols is affected more than that of our hybrid protocol as the asymmetric security parameter T grows faster than the symmetric one t (cf. Sect. 2.1.1.2).

Computation Complexity. In all Minimum protocols presented before, the M Paillier ciphertexts need to be processed, where the workload in our hybrid protocols is smaller than that of the HE-based ones due to packing. Besides the Paillier operations the following computations are needed.

In both HE-based Minimum protocols [21, 76], both C and S need to perform $\mathcal{O}(\ell' M)$ public key operations in the underlying homomorphic cryptosystem (DGK respectively EC-ElGamal) during the online phase.

When pre-computations are possible, the online phase of our hybrid Minimum protocols requires only C to perform $\mathcal{O}(\ell' M)$ symmetric cryptographic operations (SHA-256 evaluations), while S performs only XORs of bitstrings for online OT (cf. Sect. 2.2.3.3). In the setup phase (or additionally in the online phase when pre-computations are not possible), S performs $\mathcal{O}(\ell' M)$ symmetric cryptographic operations (SHA-256 evaluations) to generate the GC. For OT, both parties perform only $\mathcal{O}(t)$ public-key operations (EC multiplications) and $\mathcal{O}(\ell' M)$ symmetric cryptographic operations (SHA-256 evaluations) as described in Sect. 2.2.3.4. As SHA-256 can be evaluated more efficiently than public-key operations, the computation complexity of our hybrid protocols is lower than that of the HE-based protocols for sufficiently large databases ($\ell' M > t$).

Our implementation results in [190] show that, with pre-computations, the online phase of the hybrid face-recognition protocol of Sect. 5.3.3.3 (and hence also that of the further improved protocol of Sect. 5.3.4) is substantially faster than that of the HE-only protocols [21, 76] and scales well with increasing security level, even when implemented in the relatively slow Python programming language (compared to the optimized implementation of the HE-based protocols in C++). For $\ell'M = 320$, runtimes on an Intel Core 2 Duo at 2.4 GHz were 18 s [76], 16 s [21], and 8 s [190].

Chapter 6
Conclusion

6.1 Guidelines for Constructing Efficient SFE Protocols

The following guidelines summarize the contents of this book and can be used to construct highly efficient Secure Function Evaluation (SFE) protocols.

SFE using HE versus GC. Two-party SFE protocols can be constructed based on either Homomorphic Encryption (HE) or Garbled Circuits (GCs) which we have summarized in detail in Chap. 2. Each of these techniques has its respective advantages and disadvantages: Concerning communication, HE allows one to operate directly on the ciphertexts, whereas for GC-based SFE, helper information in form of the GC needs to be transferred for each operation (this can be done already in a setup phase). With regard to computation, HE requires computationally expensive public-key operations (these can be partly pre-computed) whereas GC is mostly based on faster symmetric-key operations—indeed, GC-based SFE protocols enable one to shift most of the complexity into a less time-critical setup phase and require only a constant (in the security parameter) number of public-key operations (Sect. 2.3.1.1).

Use HE or GC? Today's most efficient GC constructions allow "free XORs", i.e., secure evaluation of XOR gates requires no transfer of garbled tables and only negligible computation (Sect. 2.2.2.3). The performance of GC-based SFE protocols can be improved by exploiting free XORs, e.g., replacing costly non-XOR gates with smaller gates and some free XORs (Sect. 3.2). These optimizations result in improved circuit constructions for many standard functionalities (Sect. 3.3). Using these optimized circuits we have shown that GCs allow for more efficient secure comparison and first-price auctions (Sect. 3.4), whereas HE is suited better for secure (ℓ-bit integer) multiplication (Sect. 5.2.5.1) as public-key operations are faster than transferring and evaluating a large GC for multiplication with $\mathcal{O}(\ell^{1.6})$ gates (Sect. 3.3.2.2).

Use HW with GC. If available, Hardware (HW) can be used to enhance GC-based protocols in several ways. A tamper-proof HW token issued by the server can be used to locally generate GCs for the client which completely eliminates the need to transfer GCs over the Internet (Sect. 4.1). A similar token can be used for secure outsourcing of computations and arbitrary computations in a cloud computing scenario where fast response times are needed (Sect. 4.3). GCs can also be evaluated efficiently in HW which enables leakage-resilient One-Time Programs (OTPs) (Sect. 4.2).

Use HE and GC! The combination of both SFE techniques, HE and GC, in our modular SFE framework allows one to construct highly efficient SFE protocols that combine the advantages of both techniques (Sect. 5.1). Our novel SFE compiler, called Tool for Automating Secure Two-party Computations (TASTY) (Sect. 5.2), implements this framework and makes it accessible to non-experts. We consider privacy-preserving face recognition based on Eigenfaces as a representative example application that benefits from the combination of HE and GC (Sect. 5.3).

6.2 Directions for Future Research

Finally, we give some directions for future research in engineering SFE protocols for practical applications that can build upon and continue beyond the foundations laid in this book.

In the future we hope that SFE will be increasingly used in real-life applications where privacy needs to be preserved, particularly in the e-health sector. Before SFE can be used in a specific application, it needs to be investigated whether the efficiency of SFE is sufficient. On the other hand, before the functionalities to be evaluated in the specific target applications are known, it is impossible to identify the bottlenecks of current most efficient SFE constructions that need to be optimized further. This chicken-and-egg problem can be tackled with techniques from algorithm engineering [191] in a step-wise refinement process: starting from some target applications a generic tool is built with which new applications are evaluated. The encountered restrictions and bottlenecks of the tool show what aspects need to be optimized in a next version of the tool and so on. With the Tool for Automating Secure Two-party Computations (TASTY) presented in Sect. 5.2 we went through the first cycle of this process: We started with a few applications including privacy-preserving face recognition (Sect. 5.2). Meanwhile, we have preliminary results including the use of TASTY for a new application—privacy-preserving classification of medical Electro Cardiogram (ECG) data [19, 20, 186].

For future applications we expect that the functionalities that need to be evaluated securely will be substantially larger and more complex. Additionally, system requirements, e.g., for mobile devices, will be more stringent than in the applications considered so far. Therefore, we focus on SFE of large functionalities (Sect. 6.2.1) and automatic partitioning of complex functionalities (Sect. 6.2.2) in the following.

6.2.1 SFE of Large Functionalities

Today's tools for GC-based SFE (e.g., [109, 154, 157, 180]) store the entire functionality and intermediate values during its evaluation in primary memory, which currently limits the size of circuits that can be evaluated to a few million gates (approximately 4 Mio. in Fairplay and 8 Mio. in TASTY, see full version of [109]). Without fundamental changes to the architecture, evaluation of larger functionalities could be achieved by using virtual memory, i.e., transparently swapping into slower but substantially larger secondary memory, which, however, would have a severe impact on the performance. The possibility and associated problems of using secondary memory applies to SFE implementations in Software (SW) as well as in HW: SW implementations, run on today's PCs, have access to Gigabytes of primary memory Gigabytes of primary memory (RAM) and Terabytes of secondary memory (hard disk storage). Today's smartphones have only hundreds of Megabytes of primary memory (RAM) and a few Gigabytes of secondary (FLASH) memory. For HW implementations in embedded systems, memory access is even more strict as primary memory (registers and on-chip memory) is rare while access to secondary (off-chip) memory is relatively expensive. Indeed, memory access was the major bottleneck of our FPGA-based implementation for GC evaluation as discussed in Sect. 4.2.3.4.

Streaming. To overcome these memory restrictions, a streamlined evaluation is needed that never holds the entire functionality in memory, as follows: Firstly, the circuit can be compiled on-the-fly using a constant amount of memory as implemented in TASTY (see discussion in full version of [109]). Further, this stream of gates can be directly combined with our constant-memory GC creation technique of Sect. 4.1.4.3, and the garbled tables can be streamed directly over the network[1] to the evaluator who evaluates them on-the-fly as suggested in Sect. 4.1.4.1 and Sect. 4.2. The intermediate values during evaluation can be cached as described and used in Sect. 4.2.3. Finally, Oblivious Transfer (OT) can be extended dynamically as mentioned in [121] s.t. only a constant (in the security parameter) number of public key operations is needed for an arbitrary (and unknown in advance) number of OTs. We note, however, that some circuits cannot be streamed as their evaluation requires memory linear in the circuit size as described in Sect. 4.2.3.1.

For Secure Multi-party Computation (SMPC) based on HE, streaming was implemented already in the VIFF framework [69]. The recently proposed VMCrypt library [156] specifically aims to maximize GC streaming in the two-party setting. Meanwhile, parts of our techniques described above have been implemented in this architecture with corresponding performance improvements [112]. Additionally, our framework for modular SFE that combines the advantages of HE and GC (Sect. 5.1) could be integrated into this architecture for further speedups.

Parallelism. In addition to streaming, the computations in SFE protocols could also be performed in parallel to exploit the large variety of capabilities for multiprocessing available today. Parallelization can be done either on one system [e.g., using multi-core CPUs, general purpose Graphics Processing Units (GPUs), stream processors such as the Cell Broadband Engine Architecture, or Field-Programmable Gate Arrays (FPGAs)], or even distributed among multiple systems in a grid or

---

[1] Alternatively, when pre-computations are used, the GC can be buffered sequentially in secondary memory.

outsourced to the cloud.[2] The allocation of computing resources could even be performed dynamically during runtime depending on the system load (e.g., dynamic allocation of multi-core CPUs and GPUs in a grid [31]).

To allow this parallelization, it needs to be investigated how to partition the evaluated function and how to allocate sub-computations to the different computing devices efficiently.

6.2.2 Automatic Partitioning into Hybrid SFE Protocols

As soon as the evaluated functions grow larger and more complex, it will be infeasible to manually find a good partition into sub-functionalities that can be evaluated securely in our modular SFE framework of Sect. 5.1. Instead, the functionality to be evaluated should be input in a function description language (cf. Sect. 5.2.1.3) and tools should automatically determine a good partitioning into SFE sub-protocols, targeted for the specific system on which the protocol should be deployed. Possible optimization parameters could be the total amount of communication, availability of HW accelerators for GC evaluation (Sect. 4.2.3), trusted HW for GC creation (Sect. 4.1), parallel computing devices and memory (Sect. 6.2.1), network bandwidth, or even power consumption for mobile devices. The tool could automatically choose the most efficient SFE sub-protocols, what to compute where, and whether to use pre-computations or streaming. The partitioning could even be done adaptively depending on the current system state such as load or battery level. As a first step in this direction, a compiler could be built on top of TASTY (Sect. 5.2) that automatically compiles a high-level function description into different TASTY Input Language (TASTYL) programs. Afterwards, the most efficient protocol could be selected among the different choices using TASTY's benchmarking capabilities.

---

[2] Today's high performance computing services offered by cloud providers even include high-end GPUs, e.g., [9].

References

1. G. Aggarwal, N. Mishra, B. Pinkas, Secure computation of the kth-ranked element, in *Advances in Cryptology—EUROCRYPT'04, LNCS*, vol. 3027 (Springer, 2004), pp. 40–55
2. D. Agrawal, B. Archambeault, J.R. Rao, P. Rohatgi, The EM side-channel(s), in *Cryptographic Hardware and Embedded Systems (CHES'02), LNCS*, vol. 2523 (Springer, 2002), pp. 29–45
3. W. Aiello, Y. Ishai, O. Reingold, Priced oblivious transfer: how to sell digital goods, in *Advances in Cryptology—EUROCRYPT01, LNCS*, vol. 2045 (Springer, 2001) pp.119–135
4. M.-L. Akkar, C. Giraud, An implementation of DES and AES, secure against some attacks, in *Cryptographic Hardware and Embedded Systems (CHES'01), LNCS*, vol. 2162 (Springer, 2001), pp. 309–318
5. J. Algesheimer, C. Cachin, J. Camenisch, G. Karjoth, Cryptographic security for mobile code, in *IEEE Symposium on Security and Privacy (S&P'01)*, 2001, pp. 2–11
6. E. Allender, M.C. Loui, K.W. Regan, Complexity classes, in *Algorithms and Theory of Computation Handbook*, Chapter 27, ed. by M.J. Atallah (CRC Press, Rockville, 1999)
7. J.B. Almeida, E. Bangerter, M. Barbosa, S. Krenn, A.-R. Sadeghi, T. Schneider, A certifying compiler for zero-knowledge proofs of knowledge based on sigma-protocols, in *European Symposium on Research in Computer Security (ESORICS'10), LNCS*, 20–22 September 2010, vol. 6345 (Springer, 2010), pp. 151–167, http://eprint.iacr.org/2010/339
8. Amazon Elastic Compute Cloud (EC2), http://aws.amazon.com/ec2
9. Amazon High Performance Computing (HPC), http://aws.amazon.com/ec2/hpc-applications/
10. Amazon Simple Storage Service (S3), http://aws.amazon.com/s3
11. M.J. Atallah, K.N. Pantazopoulos, J.R. Rice, E.H. Spafford, Secure outsourcing of scientific computations. Adv. Comput. **54**, 216–272 (2001)
12. Y. Aumann, Y. Lindell, Security against covert adversaries: efficient protocols for realistic adversaries, in *Theory of Cryptography (TCC'07), LNCS*, vol. 4392 (Springer, 2007) pp. 137–156
13. S. Avidan, M. Butman, Efficient methods for privacy preserving face detection, in *Advances in Neural Information Processing Systems (NIPS'06)* (MIT Press, 2006), pp. 57–64
14. E. Bangerter, M. Barbosa, D.J. Bernstein, I. Damgård, D. Page, J.I. Pagter, A.-R. Sadeghi, S. Sovio, Using compilers to enhance cryptographic product development, in *Information Security Solutions, Europe (ISSE'09)*, (Vieweg+Teubner, 2009), pp. 291–301
15. E. Bangerter, S. Barzan, S. Krenn, A.-R. Sadeghi, T. Schneider, J.-K. Tsay, Bringing zero-knowledge proofs of knowledge to practice, in *International Workshop on Security Protocols (SPW'09)*, 1–3 April 2009, http://eprint.iacr.org/2009/211
16. E. Bangerter, J. Camenisch, S. Krenn, A.-R. Sadeghi, T. Schneider, Automatic generation of sound zero-knowledge protocols. *Advances in Cryptology—EUROCRYPT 2009 Poster Session*, 26–30 April 2009, http://eprint.iacr.org/2008/471

17. E. Bangerter, T. Briner, W. Henecka, S. Krenn, A.-R. Sadeghi, T. Schneider, Automatic generation of sigma-protocols, in *European Workshop on Public Key Services, Applications and Infrastructures (EUROPKI'09), LNCS*, vol. 6391 Springer, 10–11 Sept 2009, pp. 67–82, http://www.cace-project.eu/index.php?Itemid=15

18. E. Bangerter, S. Krenn, A.-R. Sadeghi, T. Schneider, YAZKC: Yet another zero-knowledge compiler. 19th USENIX Security Symposium (Security'10) Poster Session, 11–13 Aug 2010

19. M. Barni, P. Failla, V. Kolesnikov, R. Lazzeretti, A.-R. Sadeghi, T. Schneider, Secure evaluation of private linear branching programs with medical applications, in *European Symposium on Research in Computer Security (ESORICS'09), LNCS*, vol. 5789 Springer, 21–25 Sept 2009, pp. 424–439, http://eprint.iacr.org/2009/195

20. M. Barni, P. Failla, V. Kolesnikov, R. Lazzeretti, A. Paus, A.-R. Sadeghi, T. Schneider, Efficient privacy-preserving classification of ECG signals, in *IEEE International Workshop on Information Forensics and Security (IEEE WIFS'09)*, 6–9 Dec 2009, pp. 91–95

21. M. Barni, T. Bianchi, D. Catalano, M. Di Raimondo, R.D. Labati, P. Failla, D. Fiore, R. Lazzeretti, V. Piuri, F. Scotti, A. Piva, Privacy-preserving fingercode authentication, in *ACM Workshop on Multimedia and Security (MM&Sec'10) ACM*, (ACM, 2010), pp. 231–240, http://www.dmi.unict.it/diraimondo/uploads/papers/fingercode-protocol.pdf

22. M. Barni, P. Failla, R. Lazzeretti, A.-R. Sadeghi, T. Schneider, Privacy-preserving ECG classification with branching programs and neural networks. IEEE Trans. Inf. Forensics Secur. (TIFS) **6**(2), 452–468 (2011)

23. D. Beaver, Precomputing oblivious transfer, in *Advances in Cryptology—CRYPTO'95, LNCS*, vol. 963 (Springer, 1995), pp. 97–109

24. M. Bellare, P. Rogaway. Random oracles are practical: a paradigm for designing efficient protocols, in *ACM Conference on Computer and Communications Security (CCS'93)*, (ACM, 1993), pp. 62–73

25. A. Ben-David, N. Nisan, B. Pinkas, FairplayMP: a system for secure multi-party computation, in *ACM Conference on Computer and Communications Security (CCS08)*, (ACM, 2008), pp.257–266, http://fairplayproject.net/fairplayMP.html

26. M. Ben-Or, S. Goldwasser, A. Wigderson. Completeness theorems for non-cryptographic fault-tolerant distributed computation, in *ACM Symposium on Theory of Computing (STOC'88)*, (ACM, 1988), pp. 1–10

27. S. Berger, R. Cáceres, K.A. Goldman, R. Perez, R. Sailer, L. van. Doorn. vTPM: virtualizing the trusted platform module, in *USENIX Security Symposium (Security'06)*, 2006, pp. 305–320

28. C.L. Berman, Circuit width, register allocation, and ordered binary decision diagrams. IEEE Trans. CAD Integr. Circuits Syst. **10**(8), 1059–1066 (1991)

29. I.F. Blake, V. Kolesnikov, Strong conditional oblivious transfer and computing on intervals, in *Advances in Cryptology—ASIACRYPT'04, LNCS*, vol. 3329 (Springer, 2004), pp. 515–529

30. M. Blanton, P. Gasti, Secure and efficient protocols for iris and fingerprint identification, in *European Symposium on Research in Computer Security (ESORICS'11), LNCS*, vol. 6879 (Springer, 2011), pp. 190–209

31. T. Blass, *Multi-GPU Cluster use for Java/OpenMP*, Master's thesis, Friedrich-Alexander University Erlangen-Nürnberg, Germany, 2010

32. D. Bogdanov, S. Laur, J. Willemson, Sharemind: a framework for fast privacy-preserving computations, in *European Symposium on Research in Computer Security (ESORICS'08), LNCS*, vol. 5283 (Springer, 2008), pp. 192–206

33. B. Bollig, I. Wegener, Improving the variable ordering of OBDDs is NP-complete. IEEE Trans. Comput. **45**(9), 993–1002 (1996)

34. D. Boneh, E.-J. Goh, K. Nissim, Evaluating 2-DNF formulas on ciphertexts, in *Theory of Cryptography Conference (TCC'05), LNCS*, vol. 3378 (Springer, 2005), pp. 325–341

35. O. Bowcott, Interpol wants facial recognition database to catch suspects. Guardian 20 Oct 2008, http://www.guardian.co.uk/world/2008/oct/20/interpol-facial-recognition

36. J. Boyar, R. Peralta, Short discrete proofs, in Advances in Cryptology EUROCRYPT'96, LNCS, vol. 1070 (Springer, 1996), pp. 131–142
37. J. Boyar, R. Peralta, Concrete multiplicative complexity of symmetric functions, in Mathematical Foundations of Computer Science (MFCS'06), LNCS, vol. 4162 (Springer, 2006), pp. 179–189
38. J. Boyar, R. Peralta, A new combinational logic minimization technique with applications to cryptology, in Symposium on Experimental Algorithms (SOA'10), LNCS, vol. 6049 (Springer, 2010), pp. 178–189
39. J. Boyar, G. Brassard, R. Peralta, Subquadratic zero-knowledge, in IEEE Symposium on Foundations of Computer Science (FOCS'91), 1991, pp. 69–78
40. J. Boyar, C. Lund, R. Peralta, On the communication complexity of zero-knowledge proofs. J. Cryptol. 6(2), 65–85 (1993)
41. J. Boyar, G. Brassard, R. Peralta, Subquadratic zero-knowledge. J. ACM 42(6), 1169–1193 (1995)
42. J. Boyar, I. Damgård, R. Peralta, Short non-interactive cryptographic proofs. J. Cryptol. 13(4), 449–472 (2000)
43. J. Boyar, R. Peralta, D. Pochuev, On the multiplicative complexity of Boolean functions over the basis $(\wedge, \oplus, 1)$. Theor. Comput. Sci. 235(1), 43–57 (2000)
44. X. Boyen, Y. Dodis, J. Katz, R. Ostrovsky, A. Smith, Secure remote authentication using biometric data, in Advances in Cryptology EUROCRYPT'05, LNCS, vol. 3494 (Springer, 2005), pp. 147–163
45. G. Brassard, C. Crépeau, Zero-knowledge simulation of boolean circuits, in Advances in Cryptology—CRYPTO'86, LNCS, vol. 263 (Springer, 1986), pp. 223–233
46. J. Brickell, V. Shmatikov, Privacy-preserving classifier learning, in Financial Cryptography and Data Security (FC'09), LNCS, vol. 5628 (Springer, 2009), pp. 128–147
47. J. Brickell, D.E. Porter, V. Shmatikov, E. Witchel, Privacy-preserving remote diagnostics, in ACM Computer and Communications Security (CCS'07), (ACM, 2007), pp. 498–507
48. T. Briner, Compiler for zero-knowledge proof-of-knowledge protocols, Master's thesis, ETH Zurich, Switzerland, 2004
49. R.E. Bryant, On the complexity of VLSI implementations and graph representations of boolean functions with application to integer multiplication. IEEE Trans. Compu. 40(2), 205213 (1991)
50. S. Bugiel, S. Nürnberger, A.-R. Sadeghi, T. Schneider, Twin clouds: secure cloud computing with low latency, in Communications and Multimedia Security Conference (CMS11), LNCS, vol. 7025 Springer, 19–21 Oct. 2011
51. A. Bussani, J.L. Griffin, B. Jasen, K. Julisch, G. Karjoth, H. Maruyama, M. Nakamura, R. Perez, M. Schunter, A. Tanner, L. van Doorn, E. van Herreweghen, M. Waidner, S. Yoshihama, Trusted Virtual Domains: Secure Foundations for Business and IT Services. Technical Report Research Report RC23792, IBM Research, Nov 2005
52. S. Cabuk, C.I. Dalton, K. Eriksson, D. Kuhlmann, H.V. Ramasamy, G. Ramunno, A.-R. Sadeghi, M. Schunter, C. Stüble, Towards automated security policy enforcement in multi-tenant virtual data centers. J.Comput. Secur. 18, 89–121 (2010)
53. C. Cachin, J. Camenisch, J. Kilian, J. Müller. One-round secure computation and secure autonomous mobile agents, in International Colloquium on Automata, Languages and Programming (ICALP'00), LNCS, vol. 1853 (Springer, 2000), pp. 512–523
54. J. Camenisch, M. Rohe, A.-R. Sadeghi, Sokrates—a compiler framework for zero-knowledge protocols, in Western European Workshop on Research in Cryptology (WEWoRC'05), 2005
55. R. Canetti, Universally composable security: a new paradigm for cryptographic protocols, in IEEE Symposium on Foundations of Computer Science (FOCS'01), 2001, pp. 136–145
56. R. Canetti, Y. Lindell, R. Ostrovsky, A. Sahai, Universally composable two-party and multi-party secure computation, in ACM Symposium on Theory of Computing (STOC'02), (ACM, 2002), pp. 494–503

57. R. Canetti, O. Goldreich, S. Halevi, The random oracle methodology, revisited. J. ACM **51**(4), 557–594 (2004)

58. G.J. Chaitin, M.A. Auslander, A.K. Chandra, J. Cocke, M.E. Hopkins, P.W. Markstein, Register allocation via coloring. Comput. Lang. **6**(1), 47–57 (1981)

59. N. Chandran, V. Goyal, A. Sahai, New constructions for UC secure computation using tamper-proof hardware, in *Advances in Cryptology—EUROCRYPT'08, LNCS*, vol. 4965 (Springer, 2008), pp. 545–562

60. S.G. Choi, J. Katz, R. Kumaresan, H.-S Zhou, On the security of the free-xor technique. Cryptology ePrint Archive, Report 2011/510, 2011, http://eprint.iacr.org/2011/510

61. R. Chow, P. Golle, M. Jakobsson, E. Shi, J. Staddon, R. Masuoka, J. Molina, Controlling data in the cloud: outsourcing computation without outsourcing control, in *ACM Cloud Computing Security Workshop (CCSW'09)*, (ACM, 2009), pp. 85–90

62. K.-M. Chung, Y. Kalai, S. Vadhan, Improved delegation of computation using fully homomorphic encryption, in *Advances in Cryptology CRYPTO'10, LNCS*, vol. 6223 (Springer, 2010), pp. 583–501

63. Cloud Security Alliance (CSA), Top threats to cloud computing, version 1.0, http://www. cloudsecurityalliance.org/topthreats/csathreats.v1.0.pdf, March 2010

64. T.H. Cormen, C.E. Leiserson, R.L. Rivest, C. Stein, *Introduction to Algorithms*, 2nd edn. (The MIT Press, Cambridge, 2001)

65. I. Damgård, M. Jurik, A generalisation, a simplification and some applications of Paillier's probabilistic public-key system, in *Public-Key Cryptography (PKC'01), LNCS*, vol. 1992 (Springer, 2001), pp. 119–136

66. I. Damgård, M. Geisler, M. Krøigaard, Efficient and secure comparison for on-line auctions, in *Australasian Conference on Information Security and Privacy (ACISP'07), LNCS*, vol. 4586 (Springer, 2007), pp. 416–430

67. I. Damgård, M. Geisler, M. Krøigaard. A correction to Efficient and secure comparison for on-line auctions. Cryptology ePrint Archive, Report 2008/321, 2008

68. I. Damgård, M. Geisler, M. Krøigaard, Homomorphic encryption and secure comparison. J.Appl. Cryptol. **1**(1), 22–31 (2008)

69. I. Damgård, M. Geisler, M. Krøigaard, J.B. Nielsen, Asynchronous multiparty computation: theory and implementation, in *Public Key Cryptography (PKC09), LNCS*, vol. 5443 (Springer, 2009), pp. 160–179, http://viff.dk

70. I. Damgård, J.B. Nielsen, D. Wichs, Universally composable multiparty computation with partially isolated parties, in *Theory of Cryptography (TCC'09), LNCS*, vol. 5444 (Springer, 2009), pp. 315–331

71. C. Demetrescu, I. Finocchi, G.F. Italiano, Algorithm engineering, algorithmics column. Bull. EATCS **79**, 48–63 (2003)

72. N. Döttling, D. Kraschewski, J. Müller-Quade, Unconditional and composable security using a single stateful tamper-proof hardware token, in Theory of Cryptography (TCC'11), volume 6597 of LNCS, pages 164–181. Springer, 2011.

73. M. Dubovitskaya, A. Scafuro, I. Visconti, On efficient non-interactive oblivious transfer with tamper-proof hardware. Cryptology ePrint Archive, Report 2010/509, 2010, http://eprint.iacr.org/2010/509

74. ECRYPT, II, Yearly report on algorithms and keysizes (2009–2010), March 2010, http://www.ecrypt.eu.org/documents/D.SPA.13.pdf

75. T. El Gamal, A public key cryptosystem and a signature scheme based on discrete logarithms, in *Advances in Cryptology—CRYPTO'84, LNCS*, vol. 196 (Springer, 1985), pp. 10–18

76. Z. Erkin, M. Franz, J. Guajardo, S. Katzenbeisser, I. Lagendijk, T. Toft, Privacy-preserving face recognition, in *Privacy Enhancing Technologies Symposium (PETS'09), LNCS*, vol. 5672 (Springer, 2009), pp. 235–253

77. J. Feigenbaum, B. Pinkas, R.S. Ryger, F. Saint-Jean, Secure computation of surveys, in *EU Workshop on Secure Multiparty Protocols (SMP)*, ECRYPT, 2004

78. M. Fischlin, A cost-effective pay-per-multiplication comparison method for millionaires, in *Cryptographers' Track at RSA Conference (CT-RSA'01), LNCS*, vol. 2020 (Springer, 2001), pp. 457–472
79. M. Fischlin, B. Pinkas, A.-R. Sadeghi, T. Schneider, I. Visconti, Secure set intersection with untrusted hardware tokens, in *Cryptographers' Track at the RSA Conference (CT-RSA'11), LNCS*, vol. 6558 Springer, 14–18 Feb 2011, pp. 1–16
80. M. Fort, F.C. Freiling, L.D. Penso, Z. Benenson, D. Kesdogan, Trustedpals: secure multiparty computation implemented with smart cards, in *European Symposium on Research in Computer Security (ESORICS'06), LNCS*, vol. 4189 (Springer, 2006) pp. 34–48
81. M.J. Freedman, K. Nissim, B. Pinkas, Efficient private matching and set intersection, in *Advances in Cryptology EUROCRYPT'04, LNCS*, vol. 3027 (Springer, 2004), pp. 1–19
82. K.B. Frikken, Practical private DNA string searching and matching through efficient oblivious automata evaluation, in *Data and Applications Security (DBSec'09), LNCS*, vol. 5645 (Springer, 2009), pp. 81–94
83. K.B. Frikken, M.J. Atallah, J. Li, Hidden access control policies with hidden credentials, in *ACM Workshop on Privacy in the Electronic Society (WPES'04)*, (ACM, 2004), p. 27
84. K.B. Frikken, M.J. Atallah, C. Zhang, Privacy-preserving credit checking, in *ACM Conference on Electronic Commerce*, (ACM, 2005), pp. 147–154
85. K.B. Frikken, M.J. Atallah, J. Li, Attribute-based access control with hidden policies and hidden credentials. IEEE Trans. Comput. **55**(10), 1259–1270 (2006)
86. K.B. Frikken, J. Li, M.J. Atallah, Trust negotiation with hidden credentials, hidden policies, and policy cycles, in *Network and Distributed System Security Symposium (NDSS'06)*, (The Internet Society, 2006), pp. 157–172
87. K. Gandolfi, C. Mourtel, F. Olivier, Electromagnetic analysis: concrete results, in *Cryptographic Hardware and Embedded Systems (CHES'01), LNCS*, vol. 2162 (Springer, 2001), pp. 251–261
88. J.A. Garay, P. MacKenzie, K. Yang, Efficient and universally composable committed oblivious transfer and applications, in *Theory of Cryptography (TCC'04), LNCS*, vol. 2951 (Springer, 2004), pp. 297–316
89. J.A. Garay, B. Schoenmakers, J. Villegas, Practical and secure solutions for integer comparison, in *Public Key Cryptography (PKC'07), LNCS*, vol. 4450 (Springer, 2007), pp. 330–342
90. J.A. Garay, V. Kolesnikov, R. McLellan, MAC precomputation with applications to secure memory, in *Information Security Conference (ISC'09), LNCS*, vol. 5735 (Springer, 2009), pp. 427–442
91. R. Gennaro, C. Gentry, B. Parno, Non-interactive verifiable computing: outsourcing computation to untrusted workers, in *Advances in Cryptology—CRYPTO'10, LNCS*, vol. 6223 (Springer, 2010), pp. 465–482
92. C. Gentry, *A fully homomorphic encryption scheme*. Ph.D. thesis, Stanford University, 2009, http://crypto.stanford.edu/craig
93. C. Gentry, Fully homomorphic encryption using ideal lattices, in *ACM Symposium on Theory of Computing (STOC'09)*, (ACM, 2009), pp. 169–178
94. C. Gentry, S. Halevi, Implementing Gentry's fully-homomorphic encryption scheme, in *LNCS*, vol. 6632 (Springer, 2011), pp. 129–148
95. C. Gentry, S. Halevi, V. Vaikuntanathan, A simple BGN-type cryptosystem from LWE, in *Advances in Cryptology—EUROCRYPT'10, LNCS*, vol. 6110 (Springer, 2010), pp. 506–522, http://eprint.iacr.org/2010/182
96. D. Giry, J.-J. Quisquater, Cryptographic key length recommendation, 2010, http://keylength.com
97. GMP—GNU multi precision arithmetic library, http://gmplib.org
98. GMPY—multiprecision arithmetic for Python, http://code.google.com/p/gmpy
99. O. Goldreich, *Foundations of Cryptography, volume 2: Basic Applications* (Cambridge University Press, New York, 2004), http://www.wisdom.weizmann.ac.il/oded/foc-vol2.html

100. S. Goldwasser, S. Micali, Probabilistic encryption. J. Comput. Syst. Sci. **28**(2), 270–299 (1984)
101. S. Goldwasser, Y.T. Kalai, G.N. Rothblum, One-time programs, in *Advances in Cryptology—CRYPTO08, LNCS*, vol. 5157 (Springer, 2008), pp. 39–56
102. Google App Engine, https://appengine.google.com
103. V. Goyal, P. Mohassel, A. Smith, Efficient two party and multi party computation against covert adversaries, in *Advances in Cryptology—EUROCRYPT'08, LNCS*, vol. 4965 (Springer, 2008), pp. 289–306
104. V. Goyal, Y. Ishai, A. Sahai, R. Venkatesan, A. Wadia, Founding cryptography on tamper-proof hardware tokens, in *Theory of Cryptography (TCC'10), LNCS*, vol. 5978 (Springer, 2010)
105. T.K. Grose, When surveillance cameras talk. Time Magazine, 11 Feb 2008, http://www.time.com/time/world/article/0,8599,1711972,00.html
106. V. Gunupudi, S.R. Tate, Generalized non-interactive oblivious transfer using count-limited objects with applications to secure mobile agents, in *Financial Cryptography and Data Security (FC'08), LNCS*, vol. 5143 (Springer, 2008), pp. 98–112
107. J.A. Halderman, S.D. Schoen, N. Heninger, W. Clarkson, W. Paul, J.A. Calandrino, A.J. Feldman, J. Appelbaum, E.W. Felten, Lest we remember: cold boot attacks on encryption keys, in *USENIX Security Symposium (Security'08)*, 2008, pp. 45–60
108. C. Hazay, Y. Lindell, Constructions of truly practical secure protocols using standard smartcards, in *ACM Conference on Computer and Communications Security (CCS'08)*, (ACM, 2008), pp. 491–500
109. W. Henecka, S. Kögl, A.-R. Sadeghi, T. Schneider, I. Wehrenberg, TASTY: tool for automating secure two-partY computations, in *ACM Conference on Computer and Communications Security (CCS'10)*, 4–8 Oct 2010, pp. 451–462, http://eprint.iacr.org/2010/365.http://tastyproject.net
110. A. Herzberg, H. Shulman, Secure guaranteed computation. Cryptology ePrint Archive, Report 2010/449, 2010, http://eprint.iacr.org/2010/449
111. D. Hofheinz, J. Müller-Quade, D. Unruh, Universally composable zero-knowledge arguments and commitments from signature cards, in *Central European Conference on Cryptology (MoraviaCrypt'05)*, 2005
112. Y. Huang, D. Evans, J. Katz, L. Malka, Faster secure two-party computation using garbled circuits, in *USENIX Security Symposium (Security'11)*, 2011, pp. 539–554
113. Y. Huang, L. Malka, D. Evans, J. Katz, Efficient privacy-preserving biometric identification, in *Network and Distributed System Security (NDSS'11)*, (The Internet Society, 2011), http://mightbeevil.org/secure-biometrics/
114. Heise Security. Hacker extracts crypto key from TPM chip, 10 Feb 2010, http://www.h-online.com/security/news/item/Hacker-extracts-crypto-key-from-TPM-chip-927077.html
115. IBM. IBM Cryptocards, http://www-03.ibm.com/security/cryptocards/
116. A. Iliev, *Hardware-Assisted Secure Computation*. Ph.D. thesis, Dartmouth College, Hanover, NH, USA, 2009, http://www.cs.dartmouth.edu/trust/Faerieplay
117. A. Iliev, S.W. Smith, More efficient secure function evaluation using tiny trusted third parties. Technical Report TR2005-551, Dartmouth College, Computer Science, Hanover, NH, July 2005
118. A. Iliev, S.W. Smith, Faerieplay on tiny trusted third parties (work in progress), in *Workshop on Advances in Trusted, Computing (WATC'06)*, 2006
119. A. Iliev, S.W. Smith, Small, stupid, and scalable: secure computing with Faerieplay, in *ACM Workshop on Scalable Trusted Computing (STC'10)*, (ACM, 2010), pp. 41–51
120. International Civil Aviation Organization (ICAO), Machine Readable Travel Documents (MRTD), Doc 9303, Part 1, 5th edn., 2003
121. Y. Ishai, J. Kilian, K. Nissim, E. Petrank, Extending oblivious transfers efficiently, in *Advances in Cryptology—CRYPTO'03, LNCS*, vol. 2729, (Springer, 2003), pp. 145–161
122. Y. Ishai, A. Sahai, D. Wagner, Private circuits: securing hardware against probing attacks, in *Advances in Cryptology—CRYPTO'03, LNCS*, vol. 2729 (Springer, 2003), pp. 463–481

123. Y. Ishai, M. Prabhakaran, A. Sahai, Founding cryptography on oblivious transfer—efficiently, in *Advances in Cryptology—CRYPTO'08, LNCS*, vol. 5157 (Springer, 2008), pp. 572–591

124. Y. Ishai, E. Kushilevitz, R. Ostrovsky, M. Prabhakaran, A. Sahai, Efficient non-interactive secure computation, in *Advances in Cryptology—EUROCRYPT'11, LNCS*, vol. 6632 (Springer, 2011), pp. 406–425

125. S. Jarecki, V. Shmatikov, Efficient two-party secure computation on committed inputs, in *Advances in Cryptology—EUROCRYPT'07, LNCS*, vol. 4515 (Springer, 2007), pp. 97–114

126. A. Jarrous, B. Pinkas, Secure Hamming distance based computation and its applications, in *Applied Cryptography and Network Security (ACNS'09), LNCS*, vol. 5536 (Springer, 2009), pp. 107–124

127. K. Järvinen, V. Kolesnikov, A.-R. Sadeghi, T. Schneider, Efficient secure two-party computation with untrusted hardware tokens, in *Towards Hardware Intrinsic Security: Foundation and Practice, Information Security and Cryptography*, ed. by A.-R. Sadeghi, D. Naccache, (Springer-Verlag, Berlin 2010), pp. 367–386

128. K. Järvinen, V. Kolesnikov, A.-R. Sadeghi, T. Schneider, Embedded SFE: offloading server and network using hardware tokens, in *International Conference on Financial Cryptography and Data Security (FC'10), LNCS*, vol. 6052, Springer, 25–28 January 2010, pp. 207–221, http://eprint.iacr.org/2009/591

129. K. Järvinen, V. Kolesnikov, A.-R. Sadeghi, T. Schneider, Garbled circuits for leakage-resilience: hardware implementation and evaluation of one-time programs, in *International Workshop on Cryptographic Hardware and Embedded Systems (CHES10), LNCS*, vol. 6225, Springer, 1720 Aug 2010, pp. 383–397, http://eprint.iacr.org/2010/276

130. S. Jiang, S. Smith, K. Minami, Securing web servers against insider attack, in *Annual Computer Security Applications Conference (ACSAC'01), IEEE*, 2001, pp. 265–276

131. D.B. Johnson, A.J. Menezes, S. Vanstone, The elliptic curve digital signature algorithm (ECDSA). Int. J. Inf. Secur. **1**(1), 36–63 (2001)

132. V. Kabanets, J.-Y. Cai, Circuit minimization problem, in *ACM Symposium on Theory of Computing (STOC'00)*, (ACM, 2000), pp. 73–79

133. S. Kamara, K. Lauter, Cryptographic cloud storage, in *Financial Cryptography Workshops: Real-Life Cryptographic Protocols and Standardization (RLCPS'10), LNCS*, vol. 6054 (Springer, 2010), pp. 136–149

134. A.A. Karatsuba, Y. Ofman, Multiplication of many-digital numbers by automatic computers. SSSR Acad. Sci. **145**, 293–294 (1962)

135. J. Katz, Universally composable multi-party computation using tamper-proof hardware, in *Advances in Cryptology—EUROCRYPT'07, LNCS*, vol. 4515 (Springer, 2007), pp. 115–128

136. J. Katz, L. Malka, Private function evaluation with linear complexity, in *Advances in Cryptology—ASIACRYPT'11, LNCS*, vol. 7073, (Springer, 2011), pp. 556–571

137. J. Kelsey, B. Schneier, D. Wagner, C. Hall, Side channel cryptanalysis of product ciphers, in *European Sumposium on Research in Computer Security (ESORICS'98), LNCS*, vol. 1485 (Springer, 1998), pp. 97–110

138. M.S. Kiraz, B. Schoenmakers, A protocol issue for the malicious case of Yao's garbled circuit construction, in *Symposium on Information Theory in the Benelux*, 2006, pp. 283–290

139. P.C. Kocher, J. Jaffe, B. Jun, Differential power analysis, in *Advances in Cryptology—CRYPTO'99, LNCS*, vol. 1666 (Springer, 1999), pp. 388–397

140. V. Kolesnikov, Gate evaluation secret sharing and secure one-round two-party computation, in *Advances in Cryptology—ASIACRYPT'05, LNCS*, vol. 3788 (Springer, 2005), pp. 136–155

141. V. Kolesnikov, Truly efficient string oblivious transfer using resettable tamper-proof tokens, in *Theory of Cryptography Conference (TCC'10), LNCS*, vol. 5978 (Springer, 2010), pp. 327–342

142. V. Kolesnikov, T. Schneider, Improved garbled circuit: free XOR gates and applications, in *nternational Colloquium on Automata, Languages and Programming (ICALP'08), LNCS*, vol. 5126 Springer, 6–13 July 2008, pp. 486–498

143. V. Kolesnikov, T. Schneider, A practical universal circuit construction and secure evaluation of private functions, in *International Conference on Financial Cryptography and Data Security (FC'08)*, LNCS, vol. 5143, Springer, 28–31 Jan 2008, pp. 83–97, http://thomaschneider.de/FairplayPF

144. V. Kolesnikov, A.-R. Sadeghi, T. Schneider, Improved garbled circuit building blocks and applications to auctions and computing minima, in *International Conference on Cryptology and Network Security (CANS'09)*, LNCS vol. 5888, Springer, 12–14 Dec 2009, pp. 1–20, http://eprint.iacr.org/2009/411

145. V. Kolesnikov, A.-R. Sadeghi, T. Schneider, From dust to dawn: practically efficient two-party secure function evaluation protocols and their modular design. Cryptology ePrint Archive, Report 2010/079, 2010, http://eprint.iacr.org/2010/079

146. H. Krawczyk, M. Bellare, R. Canetti, HMAC: keyed-hashing for message authentication. RFC 2104 (Informational), Feb 1997, http://tools.ietf.org/html/rfc2104

147. L. Kruger, S. Jha, E.-J. Goh, D. Boneh, Secure function evaluation with ordered binary decision diagrams, in *ACM Computer and Communications Security (CCS'06)*, (ACM, 2006), pp. 410–420

148. K. Lauter, Mi. Naehrig, V. Vaikuntanathan, Can homomorphic encryption be practical?, in *ACM Cloud Computing Security Workshop (CCSW'11)*, (ACM, 2011), pp. 113–124

149. K. Lemke, Embedded security: physical protection against tampering attacks, in *Embedded Security in Cars*, Chapter 2, ed. by K. Lemke, C. Paar, M. Wolf (Springer, Berlin, 2006), pp. 207–217

150. Y. Lindell, B. Pinkas, An efficient protocol for secure two-party computation in the presence of malicious adversaries, in *Advances in Cryptology—EUROCRYPT'07, LNCS*, vol. 4515 (Springer, 2007), pp. 52–78

151. Y. Lindell, B. Pinkas, A proof of Yao's protocol for secure two-party computation. J. Cryptol. **22**(2), 161–188, 2009, http://eprint.iacr.org/2004/175

152. Y. Lindell, B. Pinkas, Secure multiparty computation for privacy-preserving data mining. J. Priv. Confid. **1**(1), 59–98 (2009)

153. Y. Lindell, B. Pinkas, Secure two-party computation via cut-and-choose oblivious transfer, in *Theory of Cryptography (TCC11), LNCS*, vol. 6597 (Springer, 2011), pp. 329–346

154. Y. Lindell, B. Pinkas, N.P. Smart, Implementing two-party computation efficiently with security against malicious adversaries, in *Security and Cryptography for Networks (SCN'08), LNCS*, vol. 5229 (Springer, 2008), pp. 2–20

155. P.D. MacKenzie, A. Oprea, M.K. Reiter, Automatic generation of two-party computations, in *ACM Conference on Computer and Communications Security (CCS'03)*, (ACM, 2003), pp. 210–219

156. L. Malka, VMCrypt—modular software architecture for scalable secure computation, in *ACM Conference on Computer and Communications Security (CCS'11)*, (ACM, 2011), pp. 715–724

157. D. Malkhi, N. Nisan, B. Pinkas, Y. Sella, Fairplay—a secure two-party computation system, in *USENIX Security Symposium (Security'04)*, 2004, pp. 287–302, http://fairplayproject.net/fairplay.html

158. S. Meiklejohn, C. Erway, A. Küpçü, T. Hinkle, A. Lysyanskaya, ZKPDL: a language-based system for efficient zero-knowledge proofs and electronic cash, in *USENIX Security Symposium (Security'10)*, 2010, pp. 193–206

159. Microsoft SQL Azure, http://www.microsoft.com/windowsazure

160. T. Moran, The Qilin crypto SDK—an open-source Java SDK for rapid prototyping of cryptographic protocols, http://qilin.seas.harvard.edu

161. T. Moran, G. Segev, David and Goliath commitments: UC computation for asymmetric parties using tamper-proof hardware, in Advances in *Cryptology—EUROCRYPT'08, LNCS*, vol. 4965 (Springer, 2008), pp. 527–544

162. I. Naumann, G. Hogben, Privacy features of European eID card specifications. Netw. Secur. **2008**(8), 9–13 (2008), (European Network and Information, Security Agency (ENISA))

163. M. Naor, B. Pinkas, Efficient oblivious transfer protocols, in *ACM-SIAM Symposium On Discrete Algorithms (SODA'01)*, (Society for Industrial and, Applied Mathematics, 2001), pp. 448–457

164. M. Naor, B. Pinkas, R. Sumner, Privacy preserving auctions and mechanism design, in *ACM Conference on Electronic Commerce*, (ACM, 1999), pp. 129–139

165. E.M. Newton, L. Sweeney, B. Malin, Preserving privacy by de-identifying face images. IEEE Trans. Knowl. Data Eng. **17**(2), 232–243 (2005)

166. J.B. Nielsen, Extending oblivious transfers efficiently—how to get robustness almost for free. Cryptology ePrint Archive, Report 2007/215, 2007, http://eprint.iacr.org/2007/215

167. J.D. Nielsen, *Languages for secure multiparty computation and towards strongly typed macros*. Ph.D. thesis, University of Aarhus, Denmark, 2009

168. J.B. Nielsen, C. Orlandi, LEGO for two-party secure computation, in *Theory of Cryptography (TCC'09)*, LNCS, vol. 5444 (Springer, 2009), pp. 368–386

169. J.D. Nielsen, M.I. Schwartzbach, A domain-specific programming language for secure multiparty computation, in *Workshop on Programming Languages and Analysis for Security (PLAS'07)*, (ACM, 2007), pp. 21–30

170. NIST, U.S. National Institute of Standards and Technology. Federal Information Processing Standards (FIPS 197). Advanced Encryption Standard (AES), Nov 2001, http://csrc.nist.gov/publications/fips/fips197/fips-197.pdf

171. NIST, U.S. National Institute of Standards and Technology. Federal Information Processing Standards (FIPS 180–2). Announcing the Secure Hash Standard, Aug 2002, http://csrc.nist.gov/publications/fips/fips180-2/fips180-2.pdf

172. J. Nzouonta, M.C. Silaghi, M. Yokoo, Secure computation for combinatorial auctions and market exchanges, in *Autonomous Agents and Multiagent Systems (AAMAS'04)*, (IEEE, 2004), pp. 1398–1399

173. M. Osadchy, B. Pinkas, A. Jarrous, B. Moskovich, SCiFI—a system for secure face identification, in *IEEE Symposium on Security and Privacy (S&P'10)*, (IEEE, 2010), pp. 239–254

174. D.A. Osvik, A. Shamir, E. Tromer, Cache attacks and countermeasures: the case of AES, in *Cryptographers' Track at—RSA Conference (CT-RSA'06)*, LNCS, vol. 3860 (Springer, 2006), pp. 1–20

175. D. Page, Theoretical use of cache memory as a cryptanalytic side-channel. Technical Report CSTR-02-003, University of Bristol, 2002

176. P. Paillier, Public-key cryptosystems based on composite degree residuosity classes, in *Advances in Cryptology—EUROCRYPT'99, LNCS*, vol. 1592 (Springer, 1999), pp. 223–238

177. A. Paus, A.-R. Sadeghi, T. Schneider, Practical secure evaluation of semi-private functions, in *International Conference on Applied Cryptography and Network Security (ACNS'09)*, LNCS, vol. 5536 Springer, 2–5 June 2009, pp. 89–106, http://www.trust.rub.de/FairplaySPF., http://eprint.iacr.org/2009/124

178. T.P. Pedersen, Non-interactive and information-theoretic secure verifiable secret sharing, in *Advances in Cryptology—CRYPTO'91, LNCS*, vol. 576 (Springer, 1992), pp. 129–140

179. K. Pietrzak, Provable security for physical cryptography, in *Western European Workshop on Research in Cryptology (WEWoRC'09)*, 2009, http://homepages.cwi.nl/pietrzak/publications/Pie09b.pdf

180. B. Pinkas, T. Schneider, N.P. Smart, S.C. Williams, Secure two-party computation is practical, in *Advances in Cryptology—ASIACRYPT 2009, LNCS*, vol. 5912 Springer, 6–10 Dec 2009, http://eprint.iacr.org/2009/314

181. Python programming language—official website, http://www.python.org

182. J.-J. Quisquater, D. Samyde, Electromagnetic analysis (EMA): measures and countermeasures for smart cards, in *Research in Smart Cards (E-smart'01), LNCS*, vol. 2140 (Springer, 2001), pp. 200–210

183. R.L. Rivest, A. Shamir, L.M. Adleman, A method for obtaining digital signatures and public-key cryptosystems. Commun. ACM **21**(2), 120–126 (1978)

184. A.-R. Sadeghi, T. Schneider, Generalized universal circuits for secure evaluation of private functions with application to data classification, in *International Conference on Information Security and Cryptology (ICISC'08), LNCS*, vol. 5461 Springer, 3–5 Dec 2008, pp. 336–353, http://eprint.iacr.org/2008/453

185. A.-R. Sadeghi, T. Schneider, Ask your e-doctor without telling: Privacy-preserving medical diagnostics. Section Days of Ruhr-University Bochum Research School, 6 Nov 2009 (Poster prize awarded)

186. A.-R. Sadeghi, T. Schneider, Verschlüsselt Rechnen: Sichere Verarbeitung verschlüsselter medizinischer Daten am Beispiel der Klassifikation von EKG-Daten, in *Workshop Innovative und sichere Informationstechnologie für das Gesundheitswesen von morgen (perspeGKtive'10), LNI*, vol. P-174 Bonner Köllen Verlag, 8 Sept 2010, pp. 11–25

187. A.-R. Sadeghi, C. Wachsmann, Trusted computing, In *Handbook of Financial Cryptography and Security* (CRC Press, 2010), pp. 221–256

188. A.-R. Sadeghi, C. Stüble, M. Winandy, Property-based TPM virtualization, in *Information Security Conference (ISC'08), LNCS*, vol. 5222 (Springer, 2008), pp. 1–16

189. A.-R. Sadeghi, T. Schneider, I. Wehrenberg, Efficient privacy-preserving face recognition, in *International Conference on Information Security and Cryptology (ICISC'09), LNCS*, vol. 5984, Springer, 2–4 Dec 2009, pp. 229–244, http://eprint.iacr.org/2009/507

190. A.-R. Sadeghi, T. Schneider, M. Winandy, Token-based cloud computing—secure outsourcing of data and arbitrary computations with lower latency, in *International Conference on Trust and Trustworthy Computing (TRUST'10)—Workshop on Trust in the Cloud, LNCS*, vol. 6101 Springer, 21–23 June 2010, pp. 417–429

191. P. Sanders, Algorithm engineering—an attempt at a definition, in *Efficient Algorithms, LNCS*, vol. 5760 (Springer, 2009), pp. 321–340

192. T. Sander, C. Tschudin, Protecting mobile agents against malicious hosts, in *Mobile Agents and Security, LNCS*, vol. 1419 (Springer, 1998), pp. 44–60

193. T. Sander, A. Young, M. Yung, Non-interactive cryptocomputing for NC^1, in *IEEE Symposium on Foundations of Computer Science (FOCS'99)*, (IEEE, 1999), pp. 554–566

194. T. Schneider, Practical secure function evaluation. Master's thesis, University Erlangen-Nürnberg, Germany, 27 Feb 2008, http://thomaschneider.de/theses/da/

195. A. Schröpfer, F. Kerschbaum, D. Biswas, S. Geißinger, C. Schütz, L1—faster development and benchmarking of cryptographic protocols, in *ECRYPT Workshop on Software Performance Enhancements for Encryption and Decryption and Cryptographic Compilers (SPEED-CC'09)*, 12–13 Oct 2009

196. A. Schröpfer, F. Kerschbaum, G. Müller, L1—an intermediate language for mixed-protocol secure computation, in *IEEE Computer Software and Applications Conference (COMPSAC'11)*, (IEEE, 2011), pp. 298–307

197. A. Shamir, How to share a secret. Commun. ACM **22**(11), 612–613 (1979)

198. C.E. Shannon, The synthesis of two-terminal switching circuits. Bell Syst. Techn. J. **28**(1), 5998 (1949)

199. M.C. Silaghi, SMC: secure multiparty computation language, http://cs.fit.edu/~msilaghi/pages/SMC/, 2004

200. S.P. Skorobogatov, Data remanence in flash memory devices, in *Cryptographic Hardware and Embedded Systems (CHES05), LNCS*, vol. 3659 (Springer, 2005), pp. 339–353

201. N.P. Smart, F. Vercauteren, Fully homomorphic encryption with relatively small key and ciphertext sizes, in *Public Key Cryptography (PKC'10), LNCS*, vol. 6056 (Springer, 2010), pp. 420–443

202. S.W. Smith, Fairy dust, secrets, and the real world. IEEE Secur. Priv. **1**(1), 89–93 (2003)

203. S.W. Smith, S. Weingart, Building a high-performance, programmable secure coprocessor. Comput. Netw. **31**(8), 831–860 (1999), (Special Issue on Computer Network Security)

204. J. Song, R. Poovendran, J. Lee, T. Iwata, The AES-CMAC Algorithm. RFC 4493 (Informational), June 2006, http://tools.ietf.org/html/rfc4493

205. Y.N. Srikant, P. Shankar (eds.) *The Compiler Design Handbook: Optimizations and Machine Code Generation* (CRC Press, New York, 2002)
206. F.-X. Standaert, O. Pereira, Y. Yu, J.-J. Quisquater, M. Yung, E. Oswald, Leakage resilient cryptography in practice. Cryptology ePrint Archive, Report 2009/341, 2009, http://eprint. iacr.org/2009/341
207. Standards for efficient cryptography, SEC 2: Recommended elliptic curve domain parameters. Technical report, Certicom Research, 2000, http://www.secg.org/download/ aid-784/sec2-v2.pdf
208. D. Stehlé, R. Steinfeld, Faster fully homomorphic encryption, in *Advances in Cryptology— ASIACRYPT'10, LNCS*, vol. 6477 (Springer, 2010), pp. 377–394
209. M. Stevens, A. Sotirov, J. Appelbaum, A.K. Lenstra, D. Molnar, D.A. Osvik, B. de Weger. Short chosen-prefix collisions for MD5 and the creation of a rogue CA certificate, in *Advances in Cryptology—CRYPTO'09, LNCS*, vol. 5677 (Springer, 2009), pp. 55–69
210. STMicroelectronics. Smartcard MCU with 32-bit ARM SecurCore SC300 CPU and 1.25 Mbytes high-density Flash memory. Data brief, Oct 2008, http://www.st.com/internet/mcu/ product/215291.jsp
211. S.R. Tate, R. Vishwanathan, Improving cut-and-choose in verifiable encryption and fair exchange protocols using trusted computing technology, in *Data and Applications Security (DBSec'09), LNCS*, vol. 5645 (Springer, 2009), pp. 252–267
212. S.R. Tate, K. Xu, Mobile agent security through multi-agent cryptographic protocols, in *International Conference on Internet Computing (IC'03)*, (CSREA Press, 2003), pp. 462–470
213. S.R. Tate, K. Xu, On garbled circuits and constant round secure function evaluation. Technical Report 2003–02, CoPS Labi, 2003
214. K. Tiri, D. Hwang, A. Hodjat, B.-C. Lai, S. Yang, P. Schaumont, I. Verbauwhede, Prototype IC with WDDL and differential routing—DPA resistance assessment, in *Cryptographic Hardware and Embedded Systems (CHES'05), LNCS*, vol. 3659 (Springer, 2005), pp. 354–365
215. J.R. Troncoso-Pastoriza, S. Katzenbeisser, M.U. Celik, Privacy preserving error resilient DNA searching through oblivious automata, in *ACM Computer and Communications Security (CCS'07)*, (ACM, 2007), pp. 519–528
216. Trusted Computing Group (TCG). TPM main specification. Main specification, Trusted Computing Group, May 2009, http://www.trustedcomputinggroup.org
217. M.A. Turk, A.P. Pentland, Eigenfaces for recognition. J. Cogn. Neurosci. **3**(1), 71–86 (1991)
218. M.A. Turk, A.P. Pentland, Face recognition using eigenfaces, in *IEEE Computer Vision and Pattern Recognition (CVPR'91)*, (IEEE, 1991), pp. 586–591
219. B.C.H. Turton, Extending quine-mccluskey for exclusive-or logic synthesis. IEEE Trans. Educat. **39**, 81–85 (1996)
220. P. Tuyls, A. H.M. Akkermans, T.A.M. Kevenaar, G.-J. Schrijen, A.M. Bazen, R.N.J. Veldhuis, Practical biometric authentication with template protection, in *Audio- and Video-Based Biometric Person Authentication, LNCS*, vol. 3546 (Springer, 2005), pp. 436–446
221. L.G. Valiant, Universal circuits (preliminary report), in *ACM Symposium on Theory of Computing (STOC'76)*, (ACM, 1976), pp. 196–203
222. M. van Dijk, A. Juels, On the impossibility of cryptography alone for privacy-preserving cloud computing, in *USENIX Workshop on Hot Topics in Security (HotSec'10)*, 2010
223. M. van Dijk, C. Gentry, S. Halevi, V. Vaikuntanathan. Fully homomorphic encryption over the integers, in *Advances in Cryptology—EUROCRYPT'10, LNCS*, vol. 6110 (Springer, 2010), pp. 24–43
224. I. Verbauwhede, P. Schaumont, Design methods for security and trust, in *Design, Automation and Test in Europe (DATE'07)*, (ACM, 2007), pp. 672–677
225. H. Vollmer, *Introduction to Circuit Complexity: A Uniform Approach* (Springer, Berlin, 1999)
226. X. Wang, Y.L. Yin, H. Yu. Finding collisions in the full SHA-1, in *Advances in Cryptology—CRYPTO'05, LNCS* vol. 3621 (Springer, 2005), pp. 17–36

227. S.H. Weingart, Physical security devices for computer subsystems: a survey of attacks and defences, in *Cryptographic Hardware and Embedded Systems (CHES'00), LNCS*, vol. 1965 (Springer, 2000), pp. 302–317

228. P. Woelfel, Bounds on the OBDD-size of integer multiplication via universal hashing. J. Comput. Syst. Sci. **71**(4), 520–534 (2005)

229. K. Xu, S.R. Tate, Universally composable secure mobile agent computation, in *Information Security Conference (ISC04), LNCS*, vol. 3225 (Springer, 2004), pp. 304–317

230. A.C. Yao, Protocols for secure computations, in *IEEE Symposium on Foundations of Computer Science (FOCS'82)*, (IEEE, 1982), pp. 160–164

231. A.C. Yao, How to generate and exchange secrets, in *IEEE Symposium on Foundations of Computer Science (FOCS'86)*, (IEEE, 1986), pp. 162–167

232. B.S. Yee, Using Secure Coprocessors, PhD thesis, School of Computer Science, Carnegie Mellon University, May 1994. CMU-CS-94-149

233. Y. Yu, J. Leiwo, B. Premkumar, Securely utilizing external computing power, in *International Symposium on Information Technology: Coding and Computing (ITCC'05)*, vol. 1 (IEEE Computer Society, 2005), pp. 762–767

234. Y. Yu, J. Leiwo, B. Premkumar, On developing privacy-preserving compilers. Int. J. Comput. Sci. Netw. Secur. (IJCSNS) **6**(3), 154–160 (2006)

Index